Student Solutions Manual

Dirk Stueber
University of Washington

Physical Chemistry for the Life Sciences

Thomas Engel
Gary Drobny
Philip Reid

Upper Saddle River, NJ 07458

Editor-in-Chief: Nicole Folchetti
Assistant Editor, Chemistry: Carol G. DuPont
Assistant Managing Editor, Science: Gina M. Cheselka
Project Manager, Science: Maureen Pancza
Supplement Cover Manager: Paul Gourhan
Supplement Cover Designer: Victoria Colotta
Operations Specialist: Amanda Smith
Director of Operations: Barbara Kittle
Cover Image: Ribbon Structure/Greg Williams

Pearson Prentice Hall
Pearson Education, Inc.
Upper Saddle River, NJ 07458

Printed in the United States of America

10 9 8 7 6 5 4 3 2 1

ISBN 13: 978-0-8053-8278-5

ISBN 10: 0-8053-8278-X

Pearson Education Ltd., *London*
Pearson Education Australia Pty. Ltd., *Sydney*
Pearson Education Singapore, Pte. Ltd.
Pearson Education North Asia Ltd., *Hong Kong*
Pearson Education Canada, Inc., *Toronto*
Pearson Educación de Mexico, S.A. de C.V.
Pearson Education—Japan, *Tokyo*
Pearson Education Malaysia, Pte. Ltd.

Table of Contents

Preface

The fundamental concepts of physical chemistry lay at the heart of research in modern physical sciences. The textbook was written with the focus on teaching the core concepts of physical chemistry, and designed to introduce the applications of these concepts to chemistry, biochemistry, and other biological sciences. The purpose of this *Selected Solutions Manual* is to assist the student in understanding the fundamental concepts and their applications. This objective is accomplished by presenting the detailed solutions to specific exemplifying problems. The selected problems cover the main concepts discussed throughout the chapters of the textbook and also expand to applications of these concepts in various fields of chemistry and biological sciences. The solutions are presented in a logical and consistent manner with the intention to illustrate good problem-solving techniques.

I have made every effort to provide accurate and instructive solutions to the selected end-of-chapter problems. If you do, however, discover errors or ambiguities in the solutions, or have suggestions for improving the instructive and educational value of the manual, please let me know by contacting me at dstueber@u.washington.edu.

Dirk Stueber

Chapter 1: Fundamental Concepts of Thermodynamics

P1.1) A sealed flask with a capacity of 1.00 dm^3 contains 5.00 g of ethane. The flask is so weak that it will burst if the pressure exceeds 1.00×10^6 Pa. At what temperature will the pressure of the gas exceed the bursting pressure?

With $pV = nRT$ and $n = \frac{m}{M}$:

$$T = \frac{p\,V\,M}{m\,R} = \frac{\left(1.00\times10^{6}\ \text{Pa}\right)\times\left(0.001\,\text{m}^{3}\right)\times\left(30.08\ \times10^{-3}\,\text{kg mol}^{-1}\right)}{\left(0.005\ \text{kg}\right)\times\left(8.314472\ \text{J K}^{-1}\ \text{mol}^{-1}\right)} = \underline{723.6\ \text{K}}$$

P1.3) Approximately how many oxygen molecules arrive each second at the mitochondrion of an active person? The following data are available: oxygen consumption is about 40. mL of O_2 per minute per kilogram of body weight, measured at $T = 300.$ K and $P = 1.0$ atm. An adult with a body weight of 64 kg has about 1×10^{12} cells. Each cell contains about 800. mitochondria.

With $pV = nRT$ the number of moles per minute and per kg of body weight is:

$$n = \frac{p\,V}{R\,T} = \frac{\left(101325\ \text{Pa}\right)\times\left(4.0\times10^{-5}\ \text{m}^{3}\right)}{\left(300\ \text{K}\right)\times\left(8.314472\ \text{J K}^{-1}\ \text{mol}^{-1}\right)} = 1.6249\times10^{-3}\ \text{mol}$$

For a body weight of 64 kg and per second the number of moles is:

$$n = \left(6.249\times10^{-3}\ \text{mol}\right)\times\left(64\ \text{kg}\right)\times\left(\frac{1\,\text{min}}{60\,\text{s}}\right) = 1.7332\times10^{-3}\ \text{mol}$$

Converting to molecules:

$$\text{molecules}\ O_2 = \left(1.7332\times10^{-3}\ \text{mol}\right)\times\left(6.02214\times10^{23}\ \text{molecules mol}^{-1}\right) = 1.04\times10^{21}\ \text{molecules}$$

With 1×10^{12} cells in a 64 kg body, and 800 mitochondria in each cell:

$$\text{molecules}\ O_2 = \frac{\left(1.04\times10^{21}\right)}{\left(1.0\times10^{12}\right)\times\left(800\right)} = \underline{1.25\times10^{6}\ \text{molecules}}$$

P1.7) A rigid vessel of volume 0.500 m^3 containing H_2 at 20.5 °C and a pressure of 611×10^3 Pa is connected to a second rigid vessel of volume 0.750 m^3 containing Ar at 31.2 °C at a pressure of 433×10^3 Pa. A valve separating the two vessels is opened and both are cooled to a temperature of 14.5 °C. What is the final pressure in the vessels?

With $pV = nRT$ the number of moles of H_2 in vessel 1 before opening the valve is:

$$n_1 = \frac{p_1\,V_1}{R\,T_1} = \frac{\left(611\times10^{3}\ \text{Pa}\right)\times\left(0.5\,\text{m}^{3}\right)}{\left(293.65\ \text{K}\right)\times\left(8.314472\ \text{J K}^{-1}\ \text{mol}^{-1}\right)} = 125.125\ \text{moles}$$

The number of moles of Ar in vessel 2 before opening the valve is:

$$n_2 = \frac{p_2 V_2}{R T_2} = \frac{(433\times10^3 \text{ Pa})\times(0.75 \text{ m}^3)}{(304.35 \text{ K})\times(8.314472 \text{ J K}^{-1} \text{ mol}^{-1})} = 128.334 \text{ moles}$$

After opening the valve the total number of moles and the total volume have to be considered:

$$p = \frac{n_{tot} R T}{V_{tot}} = \frac{(125.125 \text{ mol} + 128.334 \text{ mol})\times(8.314472 \text{ J K}^{-1} \text{ mol}^{-1})\times(287.65 \text{ K})}{(1.25\times \text{m}^3)} = \underline{4.85\times10^5 \text{ Pa}}$$

P1.10) A compressed cylinder of gas contains 1.50×10^3 g of N_2 gas at a pressure of 2.00×10^7 Pa and a temperature of 17.1 °C. What volume of gas has been released into the atmosphere if the final pressure in the cylinder is 1.80×10^5 Pa? Assume ideal behavior and that the gas temperature is unchanged.

With $pV = nRT$ and $n = \frac{m}{M}$ the volume of the cylinder is:

$$V = \frac{m R T}{p M} = \frac{(1.5 \text{ kg})\times(8.314472 \text{ J K}^{-1} \text{ mol}^{-1})\times(290.25 \text{ K})}{(2.00\times10^7 \text{ Pa})\times(0.02802 \text{ kg mol}^{-1})} = 6.4183\times10^{-3} \text{ m}^3$$

The mass of the gas left in the cylinder after the release can now be calculated using the final pressure as:

$$m = \frac{p V M}{R T} = \frac{(1.80\times10^5 \text{ Pa})\times(6.4183\times10^{-3} \text{ m}^3)\times(0.02802 \text{ kg mol}^{-1})}{(8.314472 \text{ J K}^{-1} \text{ mol}^{-1})\times(290.25 \text{ K})} = 0.0134 \text{ kg}$$

The volume of gas that was released at 1 atm can finally be obtained by using the difference in mass:

$$V = \frac{\Delta m R T}{p M} = \frac{(1.5 \text{ kg} - 0.0134 \text{ kg})\times(8.314472 \text{ J K}^{-1} \text{ mol}^{-1})\times(290.25 \text{ K})}{(101325 \text{ Pa})\times(0.02802 \text{ kg mol}^{-1})} = \underline{1.26 \text{ m}^3 = 1.26\times10^3 \text{ L}}$$

P1.13) One liter of fully oxygenated blood can carry 0.20 L of O_2 measured at $T = 273$ K and $P = 1.00$ atm. Calculate the number of moles of O_2 carried per liter of blood. Hemoglobin, the oxygen transport protein in blood, has four oxygen-binding sites. How many hemoglobin molecules are required to transport the O_2 in 1.0 L of fully oxygenated blood?

With $pV = nRT$ the number of moles of O_2 in one liter of fully oxygenated blood is:

$$n(O_2) = \frac{p V}{R T} = \frac{(101325 \text{ Pa})\times(0.0002 \text{ m}^3)}{(8.314472 \text{ J K}^{-1} \text{ mol}^{-1})\times(273 \text{ K})} = 8.9279\times10^{-3} \text{ mol}$$

Converting to molecules:

$$\text{molecules}\,O_2 = \left(8.9279\times10^{-3}\,\text{mol}\right)\times\left(6.02214\times10^{23}\ \text{molecules mol}^{-1}\right) = 5.377\times10^{21}\ \text{molecules}$$

Finally, four binding sites per hemoglobin molecule have to be considered, so that the number of O_2 molecules required is:

$$\text{molecules}\,O_2\ \text{required} = 5.377\times10^{21}\ \text{molecules}/4 = \underline{1.34\times10^{21}\ \text{molecules}}$$

P1.16) A mixture of 2.50×10^{-3} g of O_2, 3.51×10^{-3} mol of N_2, and 4.67×10^{20} molecules of CO is placed into a vessel of volume 3.50 L at 5.20 °C.

a. Calculate the total pressure in the vessel.

b. Calculate the mole fractions and partial pressures of each gas.

a) The pressure in the vessel can be calculated by using the total number of moles:

$$n(O_2) = \frac{2.5\times10^{-3}\,\text{g}}{32.0\ \text{g mol}^{-1}} = 7.8125\times10^{-5}\ \text{mol}$$

$$n(CO) = \frac{4.67\times10^{20}\ \text{molecules}}{6.02214\times10^{23}\ \text{molecules mol}^{-1}} = 7.7547\times10^{-4}\ \text{mol}$$

$$p = \frac{n_{tot}\,R\,T}{V} =$$

$$\frac{\left(7.8125\times10^{-5}\ \text{mol} + 7.7547\times10^{-4}\ \text{mol} + 3.51\times10^{-3}\right)\times\left(8.314472\ \text{J K}^{-1}\ \text{mol}^{-1}\right)\times\left(278.35\ \text{K}\right)}{\left(0.0035\ \text{m}^3\right)} =$$

$$\underline{2885.375\ \text{Pa} = 0.00288\ \text{bar}}$$

b) The mole fractions, $x_i = \frac{n_i}{n_{tot}}$, and partial pressures, $p_i = n_i \times p$, for the gases are:

$$x(O_2) = \frac{7.8125\times10^{-5}\ \text{mol}}{4.3636\times10^{-3}\ \text{mol}} = 0.0179 \qquad p(O_2) = 0.00288\ \text{bar}\times0.0179 = 5.16\times10^{-4}\ \text{bar}$$

$$x(N_2) = \frac{3.51\times10^{-3}\ \text{mol}}{4.3636\times10^{-3}\ \text{mol}} = 0.8044 \qquad p(N_2) = 0.00288\ \text{bar}\times0.8044 = 2.32\times10^{-3}\ \text{bar}$$

$$x(CO) = \frac{7.7547\times10^{-4}\ \text{mol}}{4.3636\times10^{-3}\ \text{mol}} = 0.1777 \qquad p(CO) = 0.00288\ \text{bar}\times0.1777 = 5.12\times10^{-4}\ \text{bar}$$

P1.17) Carbon monoxide (CO) competes with oxygen for binding sites on the transport protein hemoglobin. CO can be poisonous if inhaled in large quantities. A safe level of CO in air is 50. parts per million (ppm). When the CO level increases to 800. ppm, dizziness, nausea, and unconsciousness occur, followed by death. Assuming the partial pressure of oxygen in air at sea level is 0.20 atm, what ratio of O_2 to CO is fatal?

Converting the partial pressure of O_2 in the atmosphere to ppm using $x_i = \frac{p_i}{p}$:

$$x(O_2) = \frac{0.2\text{ atm}}{1\text{ atm}} = 0.2 = 20\% = 200000\text{ ppm}$$

Therefore, the fatal O_2/CO ratio is:

$$\frac{x(O_2)}{x(CO)} = \frac{200000}{800} = \underline{250}$$

P1.23) A gas sample is known to be a mixture of ethane and butane. A bulb having a 200.0-cm^3 capacity is filled with the gas to a pressure of 100.0×10^3 Pa at 20.0 °C. If the weight of the gas in the bulb is 0.3846 g, what is the mole percent of butane in the mixture?

With $pV = nRT$ the total number of moles of moles of the mixture is:

$$n_{tot} = \frac{p\,V}{R\,T} = \frac{(100 \times 10^{-3}\text{ Pa}) \times (0.0002\text{ m}^3)}{(8.314472\text{ J K}^{-1}\text{ mol}^{-1}) \times (293.15\text{ K})} = 8.2055 \times 10^{-3}\text{ mol}$$

The total number of moles can also be expressed as:

$$n_{tot} = n_{ethane} + n_{butane} = \frac{m_{ethane}}{M_{ethane}} + \frac{m_{butane}}{M_{butane}}$$

With $m_{ethane} = m_{tot} - m_{butane}$:

$$n_{tot} = \frac{((m_{tot} - m_{butane}) \times M_{butane} + m_{butane} M_{ethane})}{M_{ethane} \times M_{butane}}$$

Solving for m_{butane} and dividing by M_{butane} yields after some rearrangement:

$$n_{butane} = \frac{(n_{tot} \times M_{ethane} - m_{tot})}{(M_{ethane} - M_{butane})} = \frac{((8.2055 \times 10^{-3}\text{ mol}) \times (30.08\text{ g mol}^{-1}) - (0.3846\text{ g}))}{((30.08\text{ g mol}^{-1}) - (58.14\text{ g mol}^{-1}))} = 4.9101 \times 10^{-3}\text{ mol}$$

And finally:

$$x_{butane} = \frac{n_{butane}}{n_{total}} = \frac{(4.9101 \times 10^{-3}\text{ mol})}{(8.2055 \times 10^{-3}\text{ mol})} = \underline{0.599 = 59.9\%}$$

P1.24) A glass bulb of volume 0.136 L contains 0.7031 g of gas at 759.0 Torr and 99.5 °C. What is the molar mass of the gas?

With $M = \frac{m}{n}$ and $pV = nRT$ the molar mass of the gas is:

$$M = \frac{mRT}{pV} = \frac{(0.7031\times10^{-3}\ \text{kg})\times(8.314472\,\text{J K}^{-1}\,\text{mol}^{-1})\times(372.65\text{K})}{(101191.68\text{Pa})\times(0.000136\text{m}^3)} = \underline{0.1583\times10^{-3}\ \text{kg mol}^{-1} = 158.3\times10^{-3}\ \text{g mol}^{-1}}$$

P1.28) Calculate the pressure exerted by benzene for a molar volume of 1.42 L at 790. K using the Redlich–Kwong equation of state:

$$P = \frac{RT}{V_m - b} - \frac{a}{\sqrt{T}}\frac{1}{V_m(V_m + b)} = \frac{nRT}{V - nb} - \frac{n^2 a}{\sqrt{T}}\frac{1}{V(V + nb)}$$

The Redlich–Kwong parameters a and b for benzene are 452.0 bar dm^6 mol^{-2} K$^{1/2}$ and 0.08271 dm^3 mol^{-1}, respectively. Is the attractive or repulsive portion of the potential dominant under these conditions?

The exerted benzene pressure is calculated using $p = \frac{RT}{(V_m - b)} - \frac{a}{\sqrt{T}}\frac{1}{V_m\times(V_m + b)}$:

$$p = \frac{(8.314\times10^{-2}\ \text{L bar K}^{-1}\ \text{mol}^{-1})\times(790\ \text{K})}{(1.42\ \text{L mol}^{-1} - 0.08271\ \text{L mol}^{-1})} -$$

$$\frac{(452.0\ \text{L}^2\ \text{bar K}^{-1/2}\ \text{mol}^{-2})}{(\sqrt{790\text{K}})}\frac{1}{(1.42\ \text{L mol}^{-1})\times(1.42\ \text{L mol}^{-1} - 0.08271\ \text{L mol}^{-1})} = \underline{41.6\ \text{bar}}$$

$$M = \frac{mRT}{pV} = \frac{(0.7031\times10^{-3}\ \text{kg})\times(8.314472\,\text{J K}^{-1}\,\text{mol}^{-1})\times(372.65\text{K})}{(101191.68\text{Pa})\times(0.000136\text{m}^3)} = \underline{0.1583\times10^{-3}\ \text{kg mol}^{-1} = 158.3\times10^{-3}\ \text{g mol}^{-1}}$$

P1.31) Assume that air has a mean molar mass of 28.9 g mol^{-1} and that the atmosphere has a uniform temperature of 25.0 °C. Calculate the barometric pressure at Denver, for which z = 1600. m. Use the information contained in Problem P1.30.

Using $P_i = P_i^0 \times \text{Exp}\left[\frac{-M_i g z}{RT}\right]$ the barometric pressure at 1600 m is:

$$P_i = 1\,\text{atm}\times\text{Exp}\left[\frac{-(28.9\times10^{-3}\,\text{kg mol}^{-1})\times(9.80665\,\text{m s}^{-2})\times(1600\,\text{m})}{(8.314472\,\text{J K}^{-1}\ \text{mol}^{-1})\times(298.15\,\text{K})}\right] = \underline{0.833\,\text{atm} = 8.44\times10^4\ \text{Pa}}$$

P1.33) Aerobic cells metabolize glucose in the respiratory system. This reaction proceeds according to the overall reaction

$$6O_2(g) + C_6H_{12}O_6(s) \rightarrow 6CO_2(g) + 6H_2O(l)$$

Calculate the volume of oxygen required at STP to metabolize 0.010 kg of glucose ($C_6H_{12}O_6$). STP refers to standard temperature and pressure, that is, $T = 273$ K and $P = 1.00$ atm. Assume oxygen behaves ideally at STP.

The number of moles of O_2 according to the stoichiometry of the equation is:

$$n(O_2) = 6 \times n(\text{glucose}) = 6 \times \frac{(10\ \text{g})}{(180.18\ \text{g mol}^{-1})} = 0.0555\ \text{mol}$$

Therefore, the volume of O_2 required is:

$$V = \frac{n\,R\,T}{p} = \frac{(0.0555\ \text{mol}) \times (8.314472\ \text{J K}^{-1}\ \text{mol}^{-1}) \times (273\ \text{K})}{(101325\ \text{Pa})} = \underline{7.47 \times 10^{-3}\ \text{m}^3 = 7.47\ \text{L}}$$

P1.35) An initial step in the biosynthesis of glucose ($C_6H_{12}O_6$) is the carboxylation of pyruvic acid ($CH_3COCOOH$) to form oxaloacetic acid ($HOOCCOCH_2COOH$):

$$CH_3COCOOH(s) + CO_2(g) \rightarrow HOOCCOCH_2COOH(s)$$

If you knew nothing else about the intervening reactions involved in glucose biosynthesis other than that no further carboxylations occur, what volume of CO_2 is required to produce 0.50 g of glucose? Assume $P = 1$ atm and $T = 310.$ K.

The number of moles of CO according to the stoichiometry of the equations is:

$$n(CO_2) = n(\text{glucose}) = \frac{(0.50\ \text{g})}{(180.18\ \text{g mol}^{-1})} = 2.775 \times 10^{-3}\ \text{mol}$$

Therefore, the volume of CO_2 required is:

$$V = \frac{n\,R\,T}{p} = \frac{(2.775 \times 10^{-3}\ \text{mol}) \times (8.314472\ \text{J K}^{-1}\ \text{mol}^{-1}) \times (310\ \text{K})}{(101325\ \text{Pa})} = \underline{7.059 \times 10^{-5}\ \text{m}^3 = 0.0706\ \text{L}}$$

Chapter 2: Heat, Work, Internal Energy, Enthalpy, and the First Law of Thermodynamics

P2.1) 3.00 moles of an ideal gas at 27.0 °C expand isothermally from an initial volume of 20.0 dm^3 to a final volume of 60.0 dm^3. Calculate *w* for this process (a) for expansion against a constant external pressure of 1.00×10^5 Pa and (b) for a reversible expansion.

a) w for an expansion against a constant pressure is:

$$w = -p_{ext} \times (V_{final} - V_{initial}) = (1.00 \times 10^5 \text{ Pa}) \times (0.06 \text{ m}^3 - 0.02 \text{ m}^3) = \underline{4.00 \times 10^3 \text{ J}}$$

b) w for a reversible expansion is:

$$w = -n R T \frac{V_{final}}{V_{initial}} = -(3.00 \text{ mol}) \times (8.314472 \text{ J K}^{-1} \text{ mol}^{-1}) \times (300 \text{ K}) \times \ln \frac{(0.06 \text{ m}^3)}{(0.02 \text{ m}^3)} = \underline{-8.22 \times 10^3 \text{ J}}$$

P2.2) A major league pitcher throws a baseball at a speed of 150. km/h. If the baseball weighs 220. g and its heat capacity is 2.0 J g^{-1} K^{-1}, calculate the temperature rise of the ball when it is stopped by the catcher's mitt. Assume no heat is transferred to the catcher's mitt. Assume also that the catcher's arm does not recoil when he or she catches the ball.

The kinetic energy, $E_{kin} = \frac{1}{2} m v^2$, of the ball is transferred into heat, $q_p = C_p \Delta T$, in the mitt:

$$E_{kin} = \tfrac{1}{2} m v^2 = q_p = C_p \Delta T$$

Solving for ΔT yields:

$$\Delta T = \frac{\frac{1}{2} m v^2}{C_p} = \frac{\frac{1}{2}(0.22 \text{ kg}) \times (41.667^2 \text{ m}^2\text{s}^{-2})}{(2.0 \text{ J g}^{-1}\text{K}^{-1}) \times (220.0 \text{ g})} = \underline{0.43 \text{ K}}$$

P2.10) A muscle fiber contracts by 2.0 cm and in doing so lifts a weight. Calculate the work performed by the fiber and the weight lifted. Assume the muscle fiber obeys Hooke's law with a constant of 800. N m^{-1}.

The work is given by:

$$w = \tfrac{1}{2} k d^2 = \tfrac{1}{2}(800.0 \text{ N m}^{-1}) \times (0.02 \text{ m}^2) = \underline{0.16 \text{ J}}$$

P2.11) Calculate ΔH and ΔU for the transformation of 1.00 mol of an ideal gas from 27.0 °C and 1.00 atm to 327 °C and 17.0 atm if

$$C_{P,m} = 20.9 + 0.042\frac{T}{\mathrm{K}} \text{ in units of J K}^{-1}\text{ mol}^{-1}$$

For an ideal gas, ΔH is given by:

$$\Delta H = C_V \times \left(T_{final} - T_{initial}\right) = (20.9 + 0.042) \times (600.15K - 300.15K) = ??$$

P2.13) In the adiabatic expansion of 1.00 mol of an ideal gas from an initial temperature of 25.0 °C, the work done on the surroundings is 1200. J. If $C_{V,m} = 3/2R$, calculate q, w, ΔU, and ΔH.

For an adiabatic expansion of an ideal gas:

$$q = \underline{0}$$

$$w = \Delta U = C_{V,m} \times \left(T_{final} - T_{initial}\right) = \underline{-1200\ \mathrm{J}}$$

$$\Delta H = C_{P,m} \times \left(T_{final} - T_{initial}\right) = \left(C_{V,m} + R\right) \times \left(\frac{\Delta U}{C_{V,m}}\right) = \left(\tfrac{3}{2}R + R\right) \times \left(\frac{\Delta U}{\tfrac{3}{2}R}\right) = \tfrac{5}{3} \times \Delta U = \underline{-2000\ \mathrm{J}}$$

P2.17) Count Rumford observed that using cannon-boring machinery, a single horse could heat 11.6 kg of water (T = 273 K) to T = 355 K. in 2.5 hours. Assuming the same rate of work, how high could a horse raise a 150.-kg weight in 1 minute? Assume the heat capacity of water is 4.18 kJ K^{-1} kg^{-1}.

The rate of work, L, for warming the water sample can be calculated as the ratio of work and time:

$$L = \frac{w}{t} = \frac{C\,m\,\Delta T}{t} = \frac{\left(4.18\ \mathrm{kJ\ K^{-1}\ kg^{-1}}\right) \times (11.6\ \mathrm{kg}) \times (82\ \mathrm{K})}{(9000\ \mathrm{s})} = 0.442\ \mathrm{kJ\ s^{-1}} = 442\ \mathrm{J\ s^{-1}}$$

The rate of work for lifting the weight depends on the potential energy:

$$L = \frac{w}{t} = \frac{m\,g\,h}{t}$$, where m, g, and h are mass, gravitational acceleration, and height, respectively. Solving for h yields:

$$h = \frac{L\,t}{m\,g} = \frac{\left(442\ \mathrm{J\ s^{-1}}\right) \times (60\ \mathrm{s})}{(150.0\ \mathrm{kg}) \times \left(9.80665\ \mathrm{m\ s^{-2}}\right)} = \underline{18.03\ \mathrm{m}}$$

P2.19) 3.50 moles of an ideal gas are expanded from 450. K and an initial pressure of 5.00 bar to a final pressure of 1.00 bar, and $C_{P,m} = 5/2R$. Calculate *w* for the following two cases:

a. The expansion is isothermal and reversible.

b. The expansion is adiabatic and reversible.

Without resorting to equations, explain why the result for part (b) is greater than or less than the result for part (a).

a) Calculating the initial and final volumina:

$$V_i = \frac{n\,R\,T}{p_i} = \frac{(3.50\,\text{mol})\times(8.314472\ \text{J K}^{-1}\ \text{mol}^{-1})\times(450\,\text{K})}{(5.00\times10^5\,\text{Pa})} = 0.0262\ \text{m}^3$$

$$V_f = \frac{n\,R\,T}{p_f} = \frac{(3.50\,\text{mol})\times(8.314472\ \text{J K}^{-1}\ \text{mol}^{-1})\times(450\,\text{K})}{(1.00\times10^5\,\text{Pa})} = 0.1310\ \text{m}^3$$

w for an isothermal, reversible process is then given by:

$$w = -n\,R\,T\ln\left(\frac{V_{final}}{V_{initial}}\right) = -(3.50\,\text{mol})\times(8.314472\,\text{J K}^{-1}\ \text{mol}^{-1})\times(450\,\text{K})\times\ln\left(\frac{(0.1310\,\text{m}^3)}{(0.0262\,\text{m}^3)}\right)$$

$$w = \text{-}21076\,\text{J} = \underline{\text{-}21.1\ \text{kJ}}$$

b) For an adiabatic, reversible process:

$$\ln\left(\frac{T_{final}}{T_{initial}}\right) = -(\gamma - 1)\ \ln\left(\frac{V_{final}}{V_{initial}}\right), \text{ where } \gamma = C_{P,m}/C_{V,m}$$

$$\ln\left(\frac{T_{final}}{T_{initial}}\right) = -(\gamma - 1)\ \ln\left(\frac{T_{final}\ p_{initial}}{T_{initial}\ p_{final}}\right) = -\frac{(\gamma - 1)}{\gamma}\ \ln\left(\frac{p_{initial}}{p_{final}}\right)$$

Therefore:

$$T_{final} = \text{Exp}\left[-\frac{(\gamma - 1)}{\gamma}\ \ln\left(\frac{p_{initial}}{p_{final}}\right) + \ln(T_{initial})\right]$$

With $C_{P,m} = \frac{5}{2}R$, and $C_{V,m} = \frac{3}{2}R$, the final temperature is:

$$T_{final} = \text{Exp}\left[-0.4\times\ln\left(\frac{5\ \text{bar}}{1\ \text{bar}}\right) + \ln(450\ \text{K})\right] = 236\ \text{K}$$

And finally w for an adiabatic process and for 3.5 moles of gas:

$$w = C_V\,\Delta T = \tfrac{3}{2}\times\left(8.314472\ \mathrm{J\,K^{-1}\,mol^{-1}}\right)\times(450\,\mathrm{K} - 236.4\,\mathrm{K})\times(3.5\,\mathrm{mol}) = \underline{-9323.85\,\mathrm{J} = -9.32\,\mathrm{kJ}}$$

P2.21) 3.00 moles of an ideal gas with $C_{V,m} = 3/2R$ initially at a temperature $T_i = 298$ K and $P_i = 1.00$ bar are enclosed in an adiabatic piston and cylinder assembly. The gas is compressed by placing a 625-kg mass on the piston of diameter 20.0 cm. Calculate the work done in this process and the distance that the piston travels. Assume that the mass of the piston is negligible.

The constant pressure during compression p_{const}, which is also the final pressure, p_f, can be calculated using the gravitational acceleration. Pressure is force divided by area:

$$p_f = p_{const} = \frac{F}{A} = \frac{m\,g}{A} = \frac{m\,g}{\left(\frac{\pi}{4}d^2\right)} = \frac{(625\,\mathrm{kg})\times\left(9.80667\,\mathrm{m\,s^{-2}}\right)}{\frac{\pi}{4}(0.2\,\mathrm{m})^2} = 195084.2\,\mathrm{Pa}$$

For an adiabatic, nonreversible compression:

$$nC_{V,m}(T_f - T_i) = -p_{const}(V_f - V_i) = -p_{const}\left(\frac{n\,R\,T_f}{p_f} - \frac{n\,R\,T_i}{p_i}\right)$$

Solving for T_f and using $C_{V,m} = \tfrac{3}{2}R$ yields:

$$T_f = \frac{2}{5}\times\left(\frac{p_{const}\,T_i}{p_i} + \tfrac{3}{2}T_i\right) = \frac{2\times(195084.2\,\mathrm{Pa})\times(298\,\mathrm{K})}{5\times\left(1\times10^5\ \mathrm{Pa}\right)} + \tfrac{3}{5}(298\,\mathrm{K}) = 411.34\,\mathrm{K}$$

The work performed by the gas is:

$$w = n\,C_{V,m}(T_f - T_i) = (3.0\,\mathrm{mol})\times\tfrac{3}{2}\times\left(8.314472\ \mathrm{J\,K^{-1}\,mol^{-1}}\right)\times(411.34\,\mathrm{K} - 298\,\mathrm{K}) = \underline{4.24\times10^3\ \mathrm{J}}$$

To calculate the distance the cylinder moved, we need the final volume of the piston:

$$V_f = \frac{n\,R\,T_f}{p_f} = \frac{(3.0\,\mathrm{mol})\times\left(8.314472\ \mathrm{J\,K^{-1}\,mol^{-1}}\right)\times(411.34\,\mathrm{K})}{(195084.2\,\mathrm{Pa})} = 0.05259\,\mathrm{m}^3$$

The height of the cylinder that the gas filled initially is then:

$$h_i = \frac{V_i}{A} = \frac{\left(0.07433\,\mathrm{m}^3\right)}{\left(0.0314\,\mathrm{m}^2\right)} = 2.367\,\mathrm{m}$$

The reduced height of the cylinder filled with gas after compression is:

$$h_f = \frac{h_i\,V_f}{V_i} = \frac{(2.367\,\mathrm{m})\times\left(0.0526\,\mathrm{m}^3\right)}{\left(0.07433\,\mathrm{m}^3\right)} = 1.675\,\mathrm{m}$$

That means the piston moved the difference in height:

$$\Delta h = h_i - h_f = 2.367\,\mathrm{m} - 1.675\,\mathrm{m} = \underline{0.69\,\mathrm{m}}$$

P2.24) One mole of an ideal gas for which $C_{V,m}$ = 20.8 J K^{-1} mol^{-1} is heated from an initial temperature of 0.00 °C to a final temperature of 275 °C at constant volume. Calculate *q, w,* ΔU, and ΔH for this process.

For a process with V = constant:
w = 0

$$q = \Delta U = C_V \ \Delta T = (20.8\,\text{J K}^{-1}\ \text{mol}^{-1})\times(275\,\text{K})\times(1.0\,\text{mol}) = \underline{5720.0\,\text{J} = 5.72\,\text{kJ}}$$

$$\Delta H = C_p \ \Delta T = (n\,R + C_V)\,\Delta T =$$

$$((8.314472\,\text{J K}^{-1}\ \text{mol}^{-1})\times(1.0\,\text{mol}) + (20.8\,\text{J K}^{-1}\ \text{mol}^{-1})\times(1.0\,\text{mol}))\times(273\,\text{K}) = \underline{8006.0\,\text{J} = 8.0\,\text{kJ}}$$

P2.28) A 1.00-mol sample of an ideal gas for which $C_{V,m}$ = 3/2*R* undergoes the following two-step process: (1) From an initial state of the gas described by *T* = 28.0 °C and *P* = 2.00 × 10^4 Pa, the gas undergoes an isothermal expansion against a constant external pressure of 1.00 × 10^4 Pa until the volume has doubled. (2) Subsequently, the gas is cooled at constant volume. The temperature falls to –40.5 °C. Calculate *q, w,* ΔU, and ΔH for each step and for the overall process.

1) For an isothermal process:

$$q = w = -p_{ext}\,(V_f - V_i) = -p_{ext}\,(2\,V_i - V_i) = -p_{ext}\ V_i$$

The initial volume is:

$$V = \frac{nRT}{p} = \frac{(1\text{mol})\times(8.314472\ \text{J K}^{-1}\ \text{mol}^{-1})\times(301.15\text{K})}{(2.00\times10^4\,\text{Pa})} = 0.1251\,\text{m}^3$$

And:

$$q = w = -(0.1251\,\text{m}^3)\times(1.00\times10^4\,\text{Pa}) = \underline{-1251\,\text{J}}$$

$$\underline{\Delta H = \Delta U = 0}$$

2) For a process with V = constant:

$$\underline{w = 0}$$

$$q = \Delta U = C_V \ \Delta T = \tfrac{3}{2}(8.314472\ \text{J K}^{-1}\ \text{mol}^{-1})\times(68.5\,\text{K})\times(1.0\,\text{mol}) = \underline{854.3\,\text{J}}$$

$$\Delta H = C_p \ \Delta T = (n\,R + \tfrac{3}{2}\,n\,R)\,\Delta T = \tfrac{5}{2}\,n\,R = \tfrac{5}{2}\times(8.314472\ \text{J K}^{-1}\ \text{mol}^{-1})\times(1.0\,\text{mol})\times(68.5\,\text{K}) = \underline{1423.9\,\text{J}}$$

For the overall process:

$$w_{tot} = w_1 + w_2 = \underline{-1251\,\text{J}}$$

$$q_{tot} = q_1 + q_2 = 1251\,\text{J} + 854.3\,\text{J} = \underline{-2.1\times10^3\ \text{J}}$$

$$\Delta U_{tot} = \Delta U_1 + \Delta U_2 = \underline{854.3\,J}$$

$$\Delta H_{tot} = \Delta H_1 + \Delta H_2 = \underline{1423.9\,J}$$

P2.30) A vessel containing 1.00 mol of an ideal gas with $P_i = 1.00$ bar and $C_{P,m} = 5/2R$ is in thermal contact with a water bath. Treat the vessel, gas, and water bath as being in thermal equilibrium, initially at 298 K, and as separated by adiabatic walls from the rest of the universe. The vessel, gas, and water bath have an average heat capacity of $C_P =$ 7500. J K^{-1}. The gas is compressed reversibly to $P_f = 10.5$ bar. What is the temperature of the system after thermal equilibrium has been established?

From P2.19 for a reversible, adiabatic process:

$$T_{final} = \mathrm{Exp}\left[-\frac{(\gamma-1)}{\gamma}\ln\left(\frac{P_{initial}}{P_{final}}\right) + \ln(T_{initial})\right]$$

$$T_{final} = \mathrm{Exp}\left[-0.4\times\ln\left(\frac{1\,\text{bar}}{10.5\,\text{bar}}\right) + \ln(298\,\text{K})\right] = 763.29\,\text{K}$$

The heat transferred to the bath in the adiabatic process is:

$$q = C_{V,m}(\text{gas})\Delta T = \tfrac{5}{2}\left(8.314472\ \text{J K}^{-1}\ \text{mol}^{-1}\right)\times(763.3\,\text{K} - 298\,\text{K})\times(1.0\,\text{mol}) = 9671.6\ \text{J}$$

The temperature of the bath increases to:

$$T_f = \frac{q}{C_P(\text{system})} + T_i = \frac{(9671.6\,\text{J})}{(7500\,\text{J K}^{-1})} + (298\,\text{K}) = \underline{299.3\,\text{K}}$$

P2.31) DNA can be modeled as an elastic rod that can be twisted or bent. Suppose a DNA molecule of length L is bent such that it lies on the arc of a circle of radius R_c. The reversible work involved in bending DNA without twisting is $w_{bend} = \frac{BL}{2R^2_c}$ where B is the bending force constant. The DNA in a nucleosome particle is about 680 Å in length. Nucleosomal DNA is bent around a protein complex called the histone octamer into a circle of radius 55 Å. Calculate the reversible work involved in bending the DNA around the histone octamer if the force constant $B = 2.00 \times 10^{-28}$ J m^{-1}.

The bending work is given by:

$$w_{bend} = \frac{B\,L}{2\,R_c^2} = \frac{\left(2.00\times10^{-28}\ \text{J m}^{-1}\right)\times\left(680\times10^{-10}\ \text{m}\right)}{2\times\left(55\times10^{-10}\ \text{m}\right)^2} = \underline{2.25\times10^{-19}\ \text{J m}^{-2}}$$

P2.37) The relationship between Young's modulus and the bending force constant for a deformable cylinder is $B = EI$, where $I = \pi R^4/4$ and R is the radius of the cylinder.

a. Calculate the Young's modulus associated with a DNA of radius 10. Å (1 Å = 10^{-10} m). Assume the value of B given in Problem 2.31.

b. Suppose a DNA molecule 100 base pairs in length is extended by 10. Å. Calculate the reversible work assuming the DNA can be treated as a deformable rod.

c. Compare this work to the thermal energy. Assume T = 310. K.

a) Young's modulus is given by:

$$E = \frac{B}{I} = \frac{4\,B}{\pi\,R^4} = \frac{4\times\left(2.00\times10^{-28}\ \text{J m}^1\right)}{\pi\times\left(10\times10^{-10}\ \text{m}\right)^4} = \underline{2.55\times10^8\ \text{J m}^{-3}}$$

b) Assuming 3×10^{-10} m per base pair for the DNA, the work is given by:

$$w = \frac{E\,A_0\,\Delta L^2}{2\,L_0} = \frac{\pi\times\left(10\times10^{-10}\ \text{m}\right)^2\times\left(2.55\times10^8\ \text{J m}^{-3}\right)\times\left(10\times10^{-10}\ \text{m}\right)^2}{2\times\left(300\times10^{-10}\ \text{m}\right)} = \underline{4.45\times10^{-22}\ \text{J}}$$

c) The thermal energy at 310 K is:

$$E_{thermal} = k\,T = \left(1.38\times10^{-23}\ \text{J K}^{-1}\right)\times\left(310\ \text{K}\right) = \underline{4.28\times10^{-21}\ \text{J}}$$

P2.40) Calculate q, w, ΔU, and ΔH if 1.00 mol of an ideal gas with $C_{V,m} = 3/2R$ undergoes a reversible adiabatic expansion from an initial volume $V_i = 5.25\ \text{m}^3$ to a final volume $V_f = 25.5\ \text{m}^3$. The initial temperature is 300. K.

For an adiabatic process:

$\underline{q = 0}$

From P2.19 for a reversible, adiabatic process:

$$T_{final} = \text{Exp}\left[-(\gamma-1)\ln\left(\frac{V_{final}}{V_{initial}}\right) + \ln\left(T_{initial}\right)\right],\ \gamma = \frac{C_{p,m}}{C_{V,m}} = \frac{\left(nR + C_{V,m}\right)}{C_{V,m}} = \frac{\left(nR + \frac{3}{2}R\right)}{\frac{3}{2}R} = \frac{5}{3}$$

$$T_{final} = \text{Exp}\left[-\left(\tfrac{5}{3}-1\right)\times\ln\left(\frac{25.5\ \text{m}^3}{5.25\ \text{m}^3}\right) + \ln\left(300\ \text{K}\right)\right] = 104.6\ \text{K}$$

$$w = \Delta U = C_{V,m}\ \Delta T = \tfrac{3}{2}\times\left(8.314472\ \text{J K}^{-1}\ \text{mol}^{-1}\right)\times\left(300\ \text{K} - 104.6\ \text{K}\right)\times\left(1.0\ \text{mol}\right) = \underline{2437.0\ \text{J} = 2.44\ \text{kJ}}$$

$$\Delta H = C_{p,m}\ \Delta T = \tfrac{5}{2}\times\left(8.314472\ \text{J K}^{-1}\ \text{mol}^{-1}\right)\times\left(300\ \text{K} - 104.6\ \text{K}\right)\times\left(1.0\ \text{mol}\right) = \underline{4061.6\ \text{J} = 4.06\ \text{kJ}}$$

P2.42) One mole of an ideal gas with $C_{V,m} = 3/2R$ is expanded adiabatically against a constant external pressure of 1.00 bar. The initial temperature and pressure are T_i = 300.

K and P_i = 25.0 bar, respectively. The final pressure is P_f = 1.00 bar. Calculate q, w, ΔU, and ΔH for the process.

For an adiabatic, irreversible process:

$\underline{q = 0}$

However, dw = - pdV, where p is the pressure of the system. Rather we have dw = - p_f dV, where p_f is the constant, external pressure that is producing the work. Since we are still assuming an ideal gas in this case, we again have dU = $nC_{V,m}dT$. Equating these expressions for dw and dU, we obtain:

$$n\,C_{V,m}\;dT = -p_f\;dV$$

Integrating on both sides yields:

$$n\,C_{V,m}\,(T_f - T_i) = -p_f\,(V_f - V_i)$$

$$C_{V,m}\,(T_f - T_i) = -\frac{R\,T_f}{V_f}(V_f - V_i) = R\,T_f\left(\frac{V_i}{V_f} - 1\right)$$

Solving for T_f results in:

$$T_f = \frac{R\,T_i p_f}{p_i\,(C_{V,m} + R)} + \frac{C_{V,m}\,T_i}{(C_{V,m} + R)} = \frac{(8.314472\ \mathrm{J\,K^{-1}\,mol^{-1}})\times(300\ \mathrm{K})\times(1\times10^5\ \mathrm{Pa})}{(25\times10^5\ \mathrm{Pa})(\frac{5}{2}\times 8.314472\ \mathrm{J\,K^{-1}\,mol^{-1}})} + \tfrac{3}{5}\times(300\ \mathrm{K}) = 184.8\ \mathrm{K}$$

For an adiabatic process:

$$w = \Delta U = C_{V,m}\,\Delta T = \tfrac{3}{2}\times(8.314472\ \mathrm{J\,K^{-1}\,mol^{-1}})\times(300\,\mathrm{K} - 184.8\,\mathrm{K})\times(1.0\,\mathrm{mol}) = \underline{1436.7\,\mathrm{J} = 1.44\,\mathrm{kJ}}$$

$$\Delta H = C_{p,m}\,\Delta T = \tfrac{5}{2}\times(8.314472\ \mathrm{J\,K^{-1}\,mol^{-1}})\times(300\,\mathrm{K} - 184.8\,\mathrm{K})\times(1.0\,\mathrm{mol}) = \underline{2394.6\,\mathrm{J} = 2.39\,\mathrm{kJ}}$$

P2.44) One mole of an ideal gas, for which $C_{V,m} = 3/2R$, is subjected to two successive changes in state: (1) From 25.0 °C and 100. × 10^3 Pa, the gas is expanded isothermally against a constant pressure of 20.0 × 10^3 Pa to twice the initial volume. (2) At the end of the previous process, the gas is cooled at constant volume from 25.0 ° to –25.0 °C. Calculate q, w, ΔU, and ΔH for each of the stages. Also calculate q, w, ΔU, and ΔH for the complete process.

1) For an isothermic expansion against a constant pressure:

$$w = -p_{ext}\times(V_{final} - V_{initial}) = -p_{ext}\frac{n\,R\,T}{p_i}$$

$$w = -(100\times10^3\ \mathrm{Pa})\times\frac{(1.0\,\mathrm{mol})\times(8.314472\ \mathrm{J\,K^{-1}\,mol^{-1}})\times(298.15\,\mathrm{K})}{(100\times10^3\ \mathrm{Pa})} = \underline{-495.8\,\mathrm{J}}$$

$\underline{\Delta H = \Delta U = 0}$, since the process is isothermal

$\underline{q = -w = 495.8 \text{ J}}$

2) For cooling at constant volume:

$\underline{w = 0}$, since V = constant

$$q = \Delta U = C_V\ \Delta T = \tfrac{3}{2}\left(8.314472\ \text{J K}^{-1}\ \text{mol}^{-1}\right)\times\left(50.0\ \text{K}\right)\times\left(1.0\ \text{mol}\right) = \underline{-623.6\ \text{J}}$$

$$\Delta H = -C_p\ \Delta T = -\left(n\,R + \tfrac{3}{2}\,n\,R\right)\Delta T = \tfrac{5}{2}\,n\,R = \tfrac{5}{2}\times\left(8.314472\ \text{J K}^{-1}\ \text{mol}^{-1}\right)\times\left(1.0\ \text{mol}\right)\times\left(50.0\ \text{K}\right) = \underline{-1039.3\ \text{J}}$$

For the overall process:

$$w_{tot} = w_1 + w_2 = \underline{-495.8\ \text{J}}$$

$$q_{tot} = q_1 + q_2 = 495.8\ \text{J} - 623.6\ \text{J} = \underline{-127.8\ \text{J}}$$

$$\Delta U_{tot} = \Delta U_1 + \Delta U_2 = \underline{623.6\ \text{J}}$$

$$\Delta H_{tot} = \Delta H_1 + \Delta H_2 = \underline{-1039.3\ \text{J}}$$

Chapter 3: The Importance of State Functions: Internal Energy and Enthalpy

P3.1) A differential $dz = f(x,y)dx + g(x,y)dy$ is exact if the integral $\int f(x, y)dx + \int g(x, y)dy$ is independent of the path. Demonstrate that the differential $dz = 2xydx + x^2dy$ is exact by integrating dz along the paths $(1,1) \rightarrow (5,1) \rightarrow (5,5)$ and $(1,1) \rightarrow (3,1) \rightarrow (3,3) \rightarrow (5,3) \rightarrow (5,5)$. The first number in each set of parentheses is the x coordinate, and the second number is the y coordinate.

First determining the integrals:

$$\int 2\,x\,y\,dx = \left[x^2\,y\right]_1^u \text{ and } \int 2\,x^2\,dy = \left[x^2\,y\right]_1^u$$

With denoting the exact integral as E, the first path yields:

(1,1) → (5,1): $$E = \left[x^2\,y\right]_{x=1}^{x=5} + \left[x^2\,y\right]_{y=1}^{y=1} = (25y - y) + (0) = 24y$$

(5,1) → (5,5): $$E = \left[x^2\,y\right]_{x=5}^{x=5} + \left[x^2\,y\right]_{y=1}^{y=5} = (0) + (5x^2 - x^2) = 4x^2$$

The sum for path 1 is: $E_{path\ 1} = \underline{24y + 4x^2}$

And the second path:

(1,1) → (3,1): $$E = \left[x^2\,y\right]_{x=1}^{x=3} + \left[x^2\,y\right]_{y=1}^{y=1} = (9y - y) + (0) = 8y$$

(3,1) → (3,3): $$E = \left[x^2\,y\right]_{x=3}^{x=3} + \left[x^2\,y\right]_{y=1}^{y=3} = (0) + (3x^2 - x^2) = 2x^2$$

(3,3) → (5,3): $$E = \left[x^2\,y\right]_{x=3}^{x=5} + \left[x^2\,y\right]_{y=3}^{y=3} = (25y - 9y) + (0) = 16y$$

(5,3) → (5,5): $$E = \left[x^2\,y\right]_{x=5}^{x=5} + \left[x^2\,y\right]_{y=3}^{y=5} = (0) + (5x^2 - 3x^2) = 2x^2$$

The sum for path 2 is: $E_{path\ 2} = 8y + 2x^2 + 16y + 2x^2 \underline{= 24y + 4x^2}$

The differential is exact.

P3.3) This problem will give you practice in using the cyclic rule. Use the ideal gas law to obtain the three functions $P = f(V,T)$, $V = g(P,T)$, and $T = h(P,V)$. Show that the cyclic rule $(\partial P/\partial V)_T\,(\partial V/\partial T)_P(\partial T/\partial P)_V = -1$ is obeyed.

The functions: $p = f(V,T) = \dfrac{n\,R\,T}{V}$, $V = f(p,T) = \dfrac{n\,R\,T}{p}$, $T = f(p,V) = \dfrac{p\,V}{n\,R}$

The cyclic rule: $\left(\frac{\partial p}{\partial V}\right)_T \times \left(\frac{\partial V}{\partial T}\right)_p \times \left(\frac{\partial T}{\partial p}\right)_V = \frac{-nRT}{V^2} \times \frac{nR}{p} \times \frac{V}{nR} = \frac{-p}{V} \times \frac{V}{p} = -1$

P3.6) Because U is a state function, $(\partial/\partial V(\partial U/\partial T)_V)_T = (\partial/\partial T(\partial U/\partial V)_T)_V$. Using this relationship, show that $(\partial C_V/\partial V)_T = 0$ for an ideal gas.

$$\left(\frac{\partial C_V}{\partial V}\right)_T = \left(\frac{\partial}{\partial V}\left(\frac{\partial U}{\partial T}\right)_V\right)_T = \left(\frac{\partial}{\partial T}\left(\frac{\partial U}{\partial V}\right)_T\right)_V = \left(\frac{\partial}{\partial T}\left(T\left(\frac{\partial p}{\partial T}\right)_V - p\right)_T\right)_V$$

$$= \left(\frac{\partial}{\partial T}\left(T\frac{nR}{V} - p\right)_T\right)_V = 0$$

P3.8) Integrate the expression $\beta = 1/V(\partial V/\partial T)_P$ assuming that β is independent of pressure. By doing so, obtain an expression for V as a function of T and β at fixed P.

Integrating $\beta = \frac{1}{V}\left(\frac{\partial V}{\partial T}\right)_p$ yields:

$$\beta\, dT = \frac{1}{V} dV$$

$$\beta \int_{T_i}^{T_f} dT = \int_{V_i}^{V_f} \frac{1}{V} dV$$

$$\beta\,(T_f - T_i) = \beta\,\Delta T = \ln \frac{V_f}{V_i}$$

P3.12) Regard the enthalpy as a function of T and P. Use the cyclic rule to obtain the expression

$$C_P = -\frac{(\partial H/\partial P)_T}{(\partial T/\partial P)_H}$$

The enthalpy as a function of T and p:

$$dH = \left(\frac{\partial H}{\partial T}\right)_p dT + \left(\frac{\partial H}{\partial p}\right)_T dp$$

Using the cyclic rule: $\left(\frac{\partial p}{\partial H}\right)_T \times \left(\frac{\partial T}{\partial p}\right)_H \times \left(\frac{\partial H}{\partial T}\right)_p = -1$, C_p can be expressed as:

$$C_p = \left(\frac{\partial H}{\partial T}\right)_p = \frac{-1}{\left(\frac{\partial p}{\partial H}\right)_T\left(\frac{\partial T}{\partial p}\right)_H} = -\frac{\left(\frac{\partial H}{\partial p}\right)_T}{\left(\frac{\partial T}{\partial p}\right)_H}$$

P3.14) Use $(\partial U/\partial V)_T = (\beta T - \kappa P)/\kappa$ to evaluate $(\partial U/\partial V)_T$ for an ideal gas.

For an ideal gas:

$$\kappa = -\frac{1}{V}\left(\frac{\partial V}{\partial p}\right)_T = -\frac{1}{V}\left(-\frac{nRT}{p^2}\right) = \frac{nRT}{p^2}$$

$$\beta = \frac{1}{V}\left(\frac{\partial V}{\partial T}\right)_p = \frac{1}{V}\left(-\frac{nR}{p}\right) = \frac{nR}{Vp}$$

Therefore:

$$\left(\frac{\partial U}{\partial V}\right)_T = \frac{(\beta T - \kappa p)}{\kappa} = \frac{\beta T}{\kappa} - p = \frac{nRT}{Vp} \times \frac{p^2}{nRT} - p = \underline{p\left(\frac{1}{V} - 1\right)}$$

P3.18) Calculate w, q, ΔH, and ΔU for the process in which 1 mol of water undergoes the transition $H_2O(l, 373\text{ K}) \rightarrow H_2O(g, 460.\text{ K})$ at 1 bar of pressure. The volume of liquid water at 373 K is $1.89 \times 10^{-5}\text{ m}^3\text{ mol}^{-1}$ and the volume of steam at 373 and 460. K is 3.03 and $3.74 \times 10^{-2}\text{ m}^3\text{ mol}^{-1}$, respectively. For steam, $C_{P,m}$ can be considered constant over the temperature interval of interest at $33.58\text{ J mol}^{-1}\text{ K}^{-1}$.

$$q = \Delta H = n\,\Delta H_{vaporation} + n\,C_{p,m}^{steam}\,\Delta T = (1\text{ mol})\times(40656\text{ J}) + (1\text{ mol})\times(33.58\text{ J mol}^{-1}\text{ K}^{-1})\times(460\text{ K} - 373\text{ K})$$

$$q = \underline{4.35\times10^4\text{ J}}$$

$$w = -p_{external}\,\Delta V = (10^{-5}\text{ Pa})\times(3.03\times10^{-2}\text{m}^3 - 1.89\times10^{-5}\text{m}^3) - (10^{-5}\text{ Pa})\times(3.74\times10^{-2}\text{m}^3 - 3.03\times10^{-2}\text{m}^3)$$

$$w = (-3028\text{ J})\times(710\text{ J}) = \underline{-3.74\times10^3\text{ J}}$$

$$\Delta U = w + q = (4.35\times10^4\text{ J}) - (3738\text{ J}) = \underline{3.998\times10^4\text{ J}}$$

P3.21) The Joule coefficient is defined by $(\partial T/\partial V)_U = 1/C_V[P - T(\partial P/\partial T)_V]$. Calculate the Joule coefficient for an ideal gas and for a van der Waals gas.

For an ideal gas:

$$p = \frac{nRT}{V}$$

$$\left(\frac{\partial T}{\partial V}\right)_U = \frac{1}{C_V}\left[p - T\left(\frac{\partial p}{\partial T}\right)_V\right] = \frac{1}{C_V}\left[p - \frac{nRT}{V}\right] = 0$$

For a van der Waals gas:

$$p = \frac{nRT}{(V - nb)} - \frac{n^2 a}{V^2}$$

$$\left(\frac{\partial T}{\partial V}\right)_U = \frac{1}{C_V}\left[p - T\left(\frac{\partial p}{\partial T}\right)_V\right] = \frac{1}{C_V}\left[p - \frac{nRT}{(V - nb)}\right]$$

P3.24) Derive an expression for the internal pressure of a gas that obeys the Bethelot equation of state,

$$P = \frac{RT}{V_m - b} - \frac{a}{TV_m^2}$$

$$\left(\frac{\partial U}{\partial V}\right)_T = T\left(\frac{\partial p}{\partial T}\right)_V - p = T\left(\frac{R}{(V_m - b)} + \frac{a}{V_m^2 T^2}\right)_V - p = \frac{RT}{(V_m - b)} + \frac{a}{V_m^2 T} - p$$

P3.27) Use the result of Problem P3.26 to show that $(\partial C_V/\partial V)_T$ for the van der Waals gas is zero.

The first and second derivatives of $p = \frac{nRT}{(V - nb)} - \frac{n^2 a}{V^2}$:

$$\left(\frac{\partial p}{\partial T}\right)_V = \frac{nR}{(V - nb)} \quad \text{and} \quad \left(\frac{\partial^2 p}{\partial T^2}\right)_V = \frac{nR}{(V - nb)} = 0$$

Therefore:

$$\left(\frac{\partial C_V}{\partial p}\right)_T = T\left(\frac{\partial^2 p}{\partial T^2}\right)_V = 0$$

P3.30) Use the relation

$$C_{P,m} - C_{V,m} = T\left(\frac{\partial V_m}{\partial T}\right)_P \left(\frac{\partial P}{\partial T}\right)_V$$

the cyclic rule, and the van der Waals equation of state to derive an equation for $C_{P,m} - C_{V,m}$ in terms of V_m, T, and the gas constants R, a, and b.

Using the van der Waals equation:

$$p = \frac{RT}{(V_m - b)} - \frac{a}{V_m^{\,2}}$$

We start by evaluating:

$$T\left(\frac{\partial p}{\partial T}\right)_V = \frac{RT}{(V_m - b)}$$

Applying the cyclic rule:

$$\left(\frac{\partial V_m}{\partial T}\right)_p = \frac{-1}{\left\{\left(\frac{\partial p}{\partial T}\right)_{V_m}\left(\frac{\partial p}{\partial V_m}\right)_T\right\}} = \frac{-1}{\left\{\left(\frac{RT}{(V_m - b)}\right)\left(\frac{2a}{V_m^{\,3}} - \frac{RT}{(V_m - b)^2}\right)\right\}}$$

Therefore:

$$C_{p,m} - C_{V,m} = \frac{-1}{\left(\frac{2a}{V_m^{\,3}} - \frac{RT}{(V_m - b)^2}\right)}$$

P3.31) Show that the expression $(\partial U/\partial V)_T = T(\partial P/\partial T)_V - P$ can be written in the form

$$\left(\frac{\partial U}{\partial V}\right)_T = T^2\left(\partial\left[\frac{P}{T}\right]\Big/\partial T\right)_V = -\left(\partial\left[\frac{P}{T}\right]\Big/\partial\left[\frac{1}{T}\right]\right)_V$$

Using the van der Waals equation:

$$T^{-2}\left(\frac{\partial(P/T)}{\partial T}\right)_V = T^{-2}\left\{\left(\frac{\partial(P\,T^{-1})}{\partial T}\right)_V\right\} = T^{-2}\left\{\left(\frac{\partial P}{\partial T}T^{-1}\right)_V + \left(\frac{\partial T^{-1}}{\partial T}p\right)_V\right\} = T^{-2}\left\{\frac{1}{T}\frac{\partial P}{\partial T} + p\left(-T^{-2}\right)\right\} = T\frac{\partial P}{\partial T} - p$$

$$-\left(\frac{\partial(P/T)}{\partial(1/T)}\right)_V = -\left\{\left(\frac{\partial(P\,T^{-1})}{\partial(1/T)}\right)_V\right\} = -\left\{\left(\frac{\partial P}{\partial(1/T)}T^{-1}\right)_V + \left(\frac{\partial T^{-1}}{\partial(1/T)}p\right)_V\right\} = -\left\{-T\frac{\partial P}{\partial T} + p\right\} = T\frac{\partial P}{\partial T} - p$$

Chapter 4: Thermochemistry

P4.2) Calculate $\Delta H^\circ_{reaction}$ and $\Delta U^\circ_{reaction}$ for the oxidation of benzene. Also calculate

$$\frac{\Delta H^\circ_{reaction} - \Delta U^\circ_{reaction}}{\Delta H^\circ_{reaction}}$$

The chemical equation for the oxidation of benzene is:

$$C_6H_6(l) + 7\tfrac{1}{2}\,O_2(g) \rightarrow 6\,CO_2(g) + 3\,H_2O(l)$$

The standard enthalpy for this reaction is:

$$\Delta H^\circ = \sum_i \nu_i\, H^\circ_{f,i} = 6\times(-393.5\text{ kJ mol}^{-1}) + 3\times(-285.8\text{ kJ mol}^{-1}) - (49.1\text{ kJ mol}^{-1}) = \underline{-3268\text{ kJ mol}^{-1}}$$

ΔU is calculated as:

$$\Delta U^\circ = \Delta H^\circ - \Delta n\, R\, T =$$

$$(-3268\times10^3\text{ J mol}^{-1}) - \{(-1.5)\times(8.314472\text{ J mol}^{-1}\text{ K}^{-1})\times(298.15\text{ K})\} = \underline{-3264\text{ kJ mol}^{-1}}$$

And:

$$\frac{\Delta H^\circ - \Delta U^\circ}{\Delta H^\circ} = \frac{(-3268\times10^3\text{ J mol}^{-1}) - (-3264\times10^3\text{ J mol}^{-1})}{(-3268\times10^3\text{ J mol}^{-1})} = \underline{0.00122}$$

P4.4) Calculate ΔH for the process $N_2(g,\ 298\text{ K}) \rightarrow N_2(g,\ 650\text{ K})$ using the temperature dependence of the heat capacities from the data tables. How large is the relative error if the molar heat capacity is assumed to be constant at its value of 298.15 K over the temperature interval?

Using the temperature dependency of Cp, $\Delta H^\circ_{reaction}$ is:

$$\Delta H^\circ_{reaction} = 0\,\text{kJ mol}^{-1} + \int_{T_{low}}^{T_{high}} C_{p,i}(T)dT$$

$$= \left(A1\,T_{high} + \tfrac{1}{2}A2^2\,T_{high}{}^2 + \tfrac{1}{3}A3^3\,T_{high}{}^3 + \tfrac{1}{4}A4^4\,T_{high}{}^4\right) - \left(A1\,T_{low} + \tfrac{1}{2}A2^2\,T_{low}{}^2 + \tfrac{1}{3}A3^3\,T_{low}{}^3 + \tfrac{1}{4}A4^4\,T_{low}{}^4\right) =$$

$$\left[\begin{Bmatrix}\left(30.81\,\text{K J K}^{-1}\,\text{mol}^{-1}\times 650\,\text{K}\right)+\tfrac{1}{2}\times\left(-0.01187\,\text{K}^{-2}\,650^2\,\text{K}^2\right)\\ +\tfrac{1}{3}\times\left(2.3968\times10^{-5}\,\text{K}^{-3}\,650^3\,\text{K}^3\right)+\tfrac{1}{4}\times\left(-1.0176\times10^{-8}\,\text{K}^{-4}650^4\,\text{K}^4\right)\end{Bmatrix} - \begin{Bmatrix}\left(30.81\,\text{K J K}^{-1}\,\text{mol}^{-1}\times 298\,\text{K}\right)+\tfrac{1}{2}\times\left(-0.01187\,\text{K}^{-2}\,298^2\,\text{K}^2\right)\\ +\tfrac{1}{3}\times\left(2.3968\times10^{-5}\,\text{K}^{-3}\,298^3\,\text{K}^3\right)+\tfrac{1}{4}\times\left(-1.0176\times10^{-8}\,\text{K}^{-4}298^4\,\text{K}^4\right)\end{Bmatrix}\right]$$

$$= \left[\left\{\left(20026.5\,\text{J K}^{-1}\,\text{mol}^{-1}\right)+(-2507.5)+(2194.1)+(-454.1)\right\}-\left\{\left(9181.4\,\text{J K}^{-1}\,\text{mol}^{-1}\right)+(-527.1)+(211.4)+(-20.1)\right\}\right]$$

$$= \left[\left\{19259\,\text{J K}^{-1}\,\text{mol}^{-1}\right\}-\left\{8845.6\,\text{J K}^{-1}\,\text{mol}^{-1}\right\}\right] = \underline{10413.4\,\text{J mol}^{-1}}$$

If a constant Cp would be assumed, $\Delta H^\circ_{reaction}$ is:

$$\Delta H^\circ_{reaction} = C_{p,m,298\,K}\,\Delta T = \left(29.13\,\text{J K}^{-1}\,\text{mol}^{-1}\right)\times(352\,\text{K}) = \underline{10253.8\,\text{J}}$$

The difference between the two results is:

$$\frac{(10413.4\,\text{J} - 10253.8\,\text{J})}{(10253.8\,\text{J})} = \underline{0.016 = 1.6\%}$$

P4.11) Use the average bond energies in Table 4.3 to estimate ΔU for the reaction $C_2H_4(g) + H_2(g) \rightarrow C_2H_6(g)$. Also calculate $\Delta U^\circ_{reaction}$ from the tabulated values of ΔH°_f for reactant and products (Appendix B, Data Tables). Calculate the percent error in estimating $\Delta U^\circ_{reaction}$ from the average bond energies for this reaction.

ΔU for the hydration of ethene ($\Delta n = -1$) is given by:

$$\Delta U^\circ = \Delta H^\circ - \Delta n\,R\,T = \sum_i v_i\,H^\circ_{f,i} - \Delta n\,R\,T$$

$$\left\{-\left(52.4\,\text{kJ mol}^{-1}\right)+\left(-84.0\,\text{kJ mol}^{-1}\right)\right\}$$

$$-\left\{(-1)\times\left(8.314472\,\text{J mol}^{-1}\,\text{K}^{-1}\right)\times(298.15\,\text{K})\right\} = -136400\,\text{J} + 2479.0\,\text{J} = \underline{-133.9\,\text{kJ mol}^{-1}}$$

Using the bond energies, ΔU is calculated to be:

$$\Delta H^\circ = \sum_i D_{i,A-B}(\text{reactants}) - \sum_i D_{i,A-B}(\text{products}) = \left\{4\times\left(411\text{kJ mol}^{-1}\right)+1\times\left(432\text{kJ mol}^{-1}\right)+1\times\left(602\text{kJ mol}^{-1}\right)\right\}-$$

$$\left\{6\times\left(411\text{kJ mol}^{-1}\right)+1\times\left(346\text{kJ mol}^{-1}\right)\right\} = \underline{-134\text{kJ mol}^{-1}}$$

The difference in the results of the two calculations of ΔU is:

$$\underline{\frac{-133.9\,\text{kJ mol}^{-1}}{-134\,\text{kJ mol}^{-1}} << 1\%}$$

P4.14) At 298 K, $\Delta H^{\circ}_{reaction} = 131.28\ \text{kJ mol}^{-1}$ for the reaction C(*graphite*) + $H_2O(g) \rightarrow$ CO(*g*) + $H_2(g)$, with $C_{P,m} = 8.53$, 33.58, 29.12, and 28.82 J K^{-1} mol^{-1} for graphite, $H_2O(g)$, CO(*g*), and $H_2(g)$, respectively. Calculate $\Delta H^{\circ}_{reaction}$ at 125°C from this information. Assume that the heat capacities are independent of temperature.

Since we assume that the heat capacities are independent of temperature, $\Delta H^{\circ}_{reaction,398.15\,K}$ at 125 ºC (398.15 K) is given by:

$$\Delta H^{\circ}_{reaction,398.15\,K} = \Delta H^{\circ}_{reaction,298\,K} + \sum_i \nu_i\, C_{p,i} \Delta T =$$

$$131.28\ \text{kJ mol}^{-1} + \{-(8.53\ \text{J mol}^{-1}) - (33.58\ \text{J mol}^{-1}) + (29.12\ \text{J mol}^{-1}) + (28.82\ \text{J mol}^{-1})\} \times (100.15\text{K}) =$$

$$131.28\ \text{kJ} + 1585.4\ \text{J mol}^{-1} = \underline{132.87\ \text{kJ mol}^{-1}}$$

P4.16) Consider the reaction $TiO_2(s)$ + 2 C(*graphite*) + 2 $Cl_2(g) \rightarrow$ 2 CO(*g*) + $TiCl_4(l)$ for which $\Delta H^{\circ}_{reaction,\,298\,K} = -80\ \text{kJ mol}^{-1}$. Given the following data at 25 °C, (a) calculate $\Delta H^{\circ}_{reaction}$ at 135.8 °C, the boiling point of $TiCl_4$, and (b) calculate ΔH°_f for $TiCl_4$ (*l*) at 25 °C:

Substance	**$TiO_2(s)$**	**$Cl_2(g)$**	**C(*graphite*)**	**CO(*g*)**	**$TiCl_4(l)$**
$\Delta H^{\circ}_f\ (\text{kJ mol}^{-1})$	–945			–110.5	
$C_{P,m}(\text{J K}^{-1}\ \text{mol}^{-1})$	55.06	33.91	8.53	29.12	145.2

Assume that the heat capacities are independent of temperature.

a) $\Delta H^{\circ}_{reaction,408.95\,K}$ at 135.8 ºC (408.95 K) is given by:

$$\Delta H^{\circ}_{reaction,408.95\,K} = \Delta H^{\circ}_{reaction,298\,K} + \sum_i \nu_i\, C_{p,i} \Delta T =$$

$$-80\ \text{kJ mol}^{-1} + \left\{ \begin{matrix} -(55.06\ \text{J mol}^{-1}) - 2\times(8.53\ \text{J mol}^{-1}) - 2\times(33.91\ \text{J mol}^{-1}) \\ +2\times(29.12\ \text{J mol}^{-1}) + (145.2\ \text{J mol}^{-1}) \end{matrix} \right\} \times (63.5\text{K}) =$$

$$-80\ \text{kJ} + 7045.3\ \text{J mol}^{-1} = \underline{-73.0\ \text{kJ mol}^{-1}}$$

b) $\Delta H^{\circ}_{reaction,298\,K}$ is given by:

$$\Delta H^{\circ}_{reaction,298\,K} = -\Delta H^{\circ}_f(TiO_2(s)) - 2\,\Delta H^{\circ}_f(C(\text{graphite})) - 2\,\Delta H^{\circ}_f(Cl_2(g)) + 2\,\Delta H^{\circ}_f(CO(g)) + \Delta H^{\circ}_f(TiCl_4(l))$$

Solving for $\Delta H^{\circ}_f(TiCl_4(l))$ yields:

$$\Delta H_f^\circ(TiCl_4(l)) = \Delta H^\circ_{reaction,298\,K} + \Delta H_f^\circ(TiO_2(s)) + 2\,\Delta H_f^\circ(C(graphite))$$
$$+ 2\,\Delta H_f^\circ(Cl_2(g)) - 2\,\Delta H_f^\circ(CO(g))$$
$$= -\left(80\text{ kJ mol}^{-1}\right) - \left(945\text{ kJ mol}^{-1}\right) + \left(0\text{ kJ mol}^{-1}\right) + \left(0\text{ kJ mol}^{-1}\right) + 2\times\left(110.5\text{ kJ mol}^{-1}\right)$$
$$= \underline{-804\text{ kJ mol}^{-1}}$$

P4.20) A sample of K(*s*) of mass 2.140 g undergoes combustion in a constant volume calorimeter. The calorimeter constant is 1849 J K^{-1}, and the measured temperature rise in the inner water bath containing 1450 g of water is 2.62 K. Calculate ΔU_f° and ΔH_f° for K_2O.

The combustion reaction of K(s):

$$2\,K(s) + \tfrac{1}{2}\,O_2(g) \rightarrow K_2O(s)$$

$\Delta U^\circ_{reaction}$ from the calorimeter measurement is given by:

$$\Delta U^\circ_{reaction} = -\frac{M(K(s))}{m(K(s))}\left(\frac{m(H_2O(s))}{M(H_2O(s))}C_{p,m,H_2O}\,\Delta T + C_{calorimeter}\,\Delta T\right) =$$

$$-(2\text{ mol})\times\frac{\left(39.1\text{ g mol}^{-1}\right)}{(2.140\text{ g})}\left(\frac{(1450\text{ g})}{\left(18.02\text{ g mol}^{-1}\right)}\times\left(75.3\text{ J K}^{-1}\text{ mol}^{-1}\right)\times(2.62\text{ K}) + \left(1849\text{ J K}^{-1}\right)\times(2.62\text{ K})\right)$$

$$= \underline{-757\text{ kJ mol}^{-1}}$$

With $\Delta n = -\tfrac{1}{2}$ for the reaction, $\Delta H^\circ_{reaction}$ amounts to:

$$\Delta H^\circ = \Delta U^\circ + \Delta n\,R\,T =$$
$$\left(-757\times10^3\text{ J mol}^{-1}\right) + \left\{\left(-\tfrac{1}{2}\right)\times\left(8.314472\text{ J mol}^{-1}\text{ K}^{-1}\right)\times(298.15\text{ K})\right\} = \underline{-758\text{ kJ mol}^{-1}}$$

P4.28) Certain yeast can degrade glucose into ethanol and carbon dioxide in a process called alcoholic fermentation according to the equation:

$$C_6H_{12}O_6(s) \rightarrow 2C_2H_5OH(l) + 2CO_2(g)$$

- **a.** Calculate the enthalpy of reaction. You can find appropriate heats of formation in the appendixes.
- **b.** Calculate the work done at constant pressure of 1 bar and $T = 298$ K per mole of glucose fermented. Assume carbon dioxide behaves ideally.
- **c.** Calculate the energy change ΔU when 1 mol of glucose ferments at $T = 298$ K and 1 bar.

a) $\Delta H^\circ_{reaction,298\,K}$ is given by:

$$\Delta H^\circ_{reaction,298\,K} = -\Delta H^\circ_f(C_6H_{12}O_6(s)) + 2\,\Delta H^\circ_f(C_2H_5OH(l)) + 2\,\Delta H^\circ_f(CO_2(g))$$
$$= \left(-1273.1\,\text{kJ mol}^{-1}\right) - 2\times\left(-277.6\,\text{kJ mol}^{-1}\right) - 2\times\left(393.5\,\text{kJ mol}^{-1}\right) = \underline{-69.1\,\text{kJ mol}^{-1}}$$

b) At p_{ext} = const., for $\Delta n = 2$, the work is for 1 mol of glucose:

$$q_{lost} = -p_{ext}\,\Delta V = -p_{ext}\,\frac{R\,T}{p_{ext}}\,\Delta n = -(2\,\text{mol})\times\left(8.314472\,\text{J K}^{-1}\,\text{mol}^{-1}\right)\times(298\,\text{K}) = -4955.4\,\text{kJ}$$

c) ΔU for the fermentation of glucose comes out to be:

$$\Delta U^\circ = \Delta H^\circ - \Delta n\,R\,T =$$
$$\left(-69.1\,\text{kJ mol}^{-1}\right) - \left\{(2\,\text{mol})\times\left(8.314472\,\text{J mol}^{-1}\,\text{K}^{-1}\right)\times(298.15\,\text{K})\right\} = \underline{-74055.4\,\text{kJ mol}^{-1}}$$

P4.31) The figure below shows a DSC scan of a solution of a T4 lysozyme mutant. From the DSC data, determine T_m, the excess heat capacity δC_P and the intrinsic and transition excess heat capacities at $T = 308$ K. In your calculations, use the extrapolated curves, shown as dashed lines in the DSC scan.

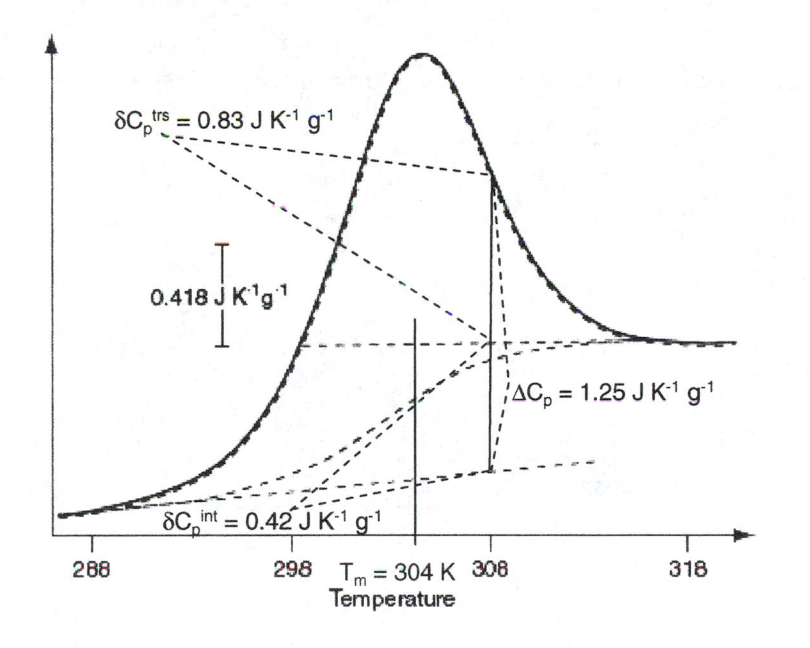

P4.35) A camper stranded in snowy weather loses heat by wind convection. The camper is packing emergency rations consisting of 60.% sucrose, 30.% fat, and 10.% protein by weight. Using the data provided in Problem P4.34 and assuming the fat content of the rations can be treated with palmitic acid data and the protein content similarly by the protein data in Problem P4.34, how much emergency rations must the camper consume to compensate for a reduction in body temperature of 4.0 K? Assume the heat capacity of the body equals that of water. Assume the camper weighs 70. kg. State any additional assumptions.

At constant pressure q = ΔH. The composition of the emergency rations means that 1 kg of the rations contains the following number of moles of sucrose, fat, and protein:

$$n_{succrose} = \frac{m}{M} = \frac{(600\,g)}{(342.3\,g\,mol^{-1})} = 1.753\,mol$$

$$n_{fat} = \frac{m}{M} = \frac{(300\,g)}{(256.43\,g\,mol^{-1})} = 1.17\,mol$$

$$n_{protein} = \frac{m}{M} = \frac{(100\,g)}{(88.30\,g\,mol^{-1})} = 1.13\,mol$$

Therefore, the enthalpy of combustion for 1 kg of rations is:

$$\Delta H^{\circ}_{combustion,1\,kg} = (1.753\,mol)\times(-5647\,kJ\,mol^{-1}) + (1.17\,mol)\times(-10035\,kJ\,mol^{-1}) + (1.13\,mol)\times(-22\,kJ\,mol^{-1})$$
$$= -21665\,kJ$$

The heat the stranded camper loses is given by:

$$q_{lost} = n_{H_2O}\,C_{p,m}\,\Delta T = \frac{m_{H_2O}}{M_{H_2O}}\,C_{p,m}\,\Delta T = \frac{(70000\,g)}{(18.01\,g\,mol^{-1})}\times(75.3\,J\,K^{-1}\,mol^{-1})\times(4.0\,K) = 1170.7\,kJ$$

Finally, the mass of rations that needs to be consumed to produce the lost amount of heat assuming the body consists of 90% water is then:

$$m_{rations} = 0.9\times\frac{(1\,kg)\times(1170.7kJ)}{(21665kJ)} = \underline{0.0486\,kg = 49\,g}$$

Chapter 5: Entropy and the Second and Third Laws of Thermodynamics

P5.2) Consider the reversible Carnot cycle shown in Figure 5.2 with 1 mol of an ideal gas with $C_V = 3/2R$ as the working substance. The initial isothermal expansion occurs at the hot reservoir temperature of $T_{hot} = 600$ °C from an initial volume of 3.50 L (V_a) to a volume of 10.0 L (V_b). The system then undergoes an adiabatic expansion until the temperature falls to $T_{cold} = 150.$°C. The system then undergoes an isothermal compression and a subsequent adiabatic compression until the initial state described by $T_a = 600.$ °C and $V_a = 3.50$ L is reached.

a. Calculate V_c and V_d.

b. Calculate w for each step in the cycle and for the total cycle.

c. Calculate ε and the amount of heat that is extracted from the hot reservoir to do 1.00 kJ of work in the surroundings.

a) V_c results from an adiabatic expansion:

$$C_V \ln\frac{T_{hot}}{T_{cold}} = -n\,R\,T\ln\frac{V_c}{V_b}$$

Solving for V_c yields:

$$V_c = V_b\ \mathrm{Exp}\left[-\tfrac{3}{2}\times\ln\frac{T_{cold}}{T_{hot}}\right] = (10.0\ \mathrm{L})\times\mathrm{Exp}\left[-\tfrac{3}{2}\times\ln\frac{(423.15\ \mathrm{K})}{(873.15\ \mathrm{K})}\right] = \underline{29.64\ \mathrm{L}}$$

V_d can be obtained by using the relationship of two isothermal processes:

$$n\,R\,T\ln\frac{V_a}{V_b} = n\,R\,T\ln\frac{V_c}{V_d}$$

$$V_d = \frac{V_b\,V_c}{V_a} = \frac{(29.64\ \mathrm{L})\times(3.5\ \mathrm{L})}{(10.0\ \mathrm{L})} = \underline{10.4\ \mathrm{L}}$$

b) The work for each of the four steps:

a → b, isothermal expansion, w = -q, $\Delta U = 0$:

$$w = -n\,R\,T_{hot}\ln\frac{V_b}{V_a} = -(1\ \mathrm{mol})\times(8.314472\ \mathrm{J\ mol^{-1}\ K^{-1}})\times(873.15\ \mathrm{K})\times\ln\frac{(10.0\ \mathrm{L})}{(3.5\ \mathrm{L})} = \underline{-7.62\ \mathrm{kJ}}$$

b → c, adiabatic expansion, q = 0, $\Delta U = w$:

$$w = C_V\,\Delta T = \tfrac{3}{2}\times(8.314472\ \mathrm{J\ mol^{-1}\ K^{-1}})\times(423.15\ \mathrm{K} - 873.15\ \mathrm{K}) = \underline{-5.61\ \mathrm{kJ}}$$

c → d, isothermal compression, w = -q, $\Delta U = 0$:

$$w = -n\,R\,T_{cold}\,\ln\frac{V_c}{V_d} = -(1\,\text{mol})\times(8.314472\,\text{J mol}^{-1}\,\text{K}^{-1})\times(423.15\,\text{K})\times\ln\frac{(10.4\,\text{L})}{(29.64\,\text{L})} = \underline{3.68\,\text{kJ}}$$

d → a, adiabatic compression, $q = 0$, $\Delta U = w$:

$$w = C_V\,\Delta T = \tfrac{3}{2}\times(8.314472\,\text{J mol}^{-1}\,\text{K}^{-1})\times(873.15\,\text{K} - 423.15\,\text{K}) = \underline{5.61\,\text{kJ}}$$

The total cycle:

$$w_{tot} = w_{a\to b} + w_{a\to b} + w_{a\to b} + w_{a\to b} = -7.62\,\text{kJ} - 5.61\,\text{kJ} + 3.68\,\text{kJ} + 5.61\,\text{kJ} = \underline{-3.94\,\text{kJ}}$$

c) $$\varepsilon = \frac{q_{ab}\;q_{cd}}{q_{ab}} = \frac{(7.62\,\text{kJ}) + (-3.68\,\text{kJ})}{(7.62\,\text{kJ})} = \underline{0.517}$$

The amount of heat extracted is:

$$q_{ext} = \frac{(1.0\;\text{kJ}) + (-7.62\,\text{kJ})}{(-3.94\,\text{kJ})} = \underline{1.93\,\text{kJ}}$$

P5.5) Calculate ΔS if the temperature of 1 mol of an ideal gas with $C_V = 3/2\,R$ is increased from 150. to 350. K under conditions of (a) constant pressure and (b) constant volume.

a) ΔS at p = constant:

$$\Delta S = C_{p,m}\,\ln\frac{T_f}{T_i} = n\,(R + C_{V,m})\ln\frac{T_f}{T_i} = \tfrac{5}{2}\,n\,R\,\ln\frac{T_f}{T_i}$$

$$= \tfrac{5}{2}\times(1\,\text{mol})\times(8.314472\,\text{J mol}^{-1}\,\text{K}^{-1})\times\ln\frac{(350\,\text{K})}{(150\,\text{K})} = \underline{17.6\,\text{J K}^{-1}}$$

b) ΔS at V = constant:

$$\Delta S = C_{V,m}\,\ln\frac{T_f}{T_i} = n\,C_{V,m}\,\ln\frac{T_f}{T_i} = \tfrac{3}{2}\,n\,R\,\ln\frac{T_f}{T_i}$$

$$= \tfrac{3}{2}\times(1\,\text{mol})\times(8.314472\,\text{J mol}^{-1}\,\text{K}^{-1})\times\ln\frac{(350\,\text{K})}{(150\,\text{K})} = \underline{10.6\,\text{J K}^{-1}}$$

P5.9) The average heat evolved by the oxidation of foodstuffs in an average adult per hour per kilogram of body weight is 7.20 kJ $\text{kg}^{-1}\,\text{h}^{-1}$. Suppose the heat evolved by this oxidation is transferred into the surroundings over a period lasting 1 day. Calculate the entropy change of the surroundings associated with this heat transfer. Assume the weight of an average adult is 70.0 kg. Assume also the surroundings are at $T = 293.0$ K.

The heat evolved over one day for a 70.0 kg person is:

$$q(1\,\text{day},70.0\,\text{kg}) = (7.20\,\text{kJ kg}^{-1}\,\text{h}^{-1})\times(24\,\text{h d}^{-1})\times(70\,\text{kg}) = 12096\,\text{kJ}$$

ΔS is then:

$$\Delta S = \frac{dq}{T} = \frac{(12096\,\text{kJ})}{(293.0\,\text{K})} = \underline{41.3\,\text{kJ K}^{-1}}$$

P5.11) At the transition temperature of 95.4 °C, the enthalpy of transition from rhombic to monoclinic sulfur is 0.38 kJ mol^{-1}.

a. Calculate the entropy of transition under these conditions.

b. At its melting point, 119 °C, the enthalpy of fusion of monoclinic sulfur is 1.23 kJ mol^{-1}. Calculate the entropy of fusion.

c. The values given in parts (a) and (b) are for 1 mol of sulfur; however, in crystalline and liquid sulfur, the molecule is present as S_8. Convert the values of the enthalpy and entropy of fusion in parts (a) and (b) to those appropriate for S_8.

a) $$\Delta S_{trans} = \frac{dq}{T} = \frac{\Delta H}{T} = \frac{(0.38\,\text{kJ mol}^{-1})}{(368.55\,\text{K})} = \underline{1.03\,\text{J K}^{-1}}$$

b) $$\Delta S_{fus} = \frac{dq}{T} = \frac{\Delta H}{T} = \frac{(1.23\,\text{kJ mol}^{-1})}{(3925\,\text{K})} = \underline{3.14\,\text{J K}^{-1}}$$

c) ΔS_{trans} and ΔS_{fus} for S_8 are:

$$\Delta S_{trans}(S_8) = 8\times(1.03\,\text{kJ mol}^{-1}) = \underline{8.24\,\text{J K}^{-1}}$$

$$\Delta S_{fus}(S_8) = 8\times(1.03\,\text{kJ mol}^{-1}) = \underline{25.12\,\text{J K}^{-1}}$$

P5.12)

a. Calculate ΔS if 1 mol of liquid water is heated from 0 °C to 100.°C under constant pressure if $C_{P,m} = 75.291$ J K^{-1} mol^{-1}.

b. The melting point of water at the pressure of interest is 0 °C and the enthalpy of fusion is 6.0095 kJ mol^{-1}. The boiling point is 100. °C and the enthalpy of vaporization is 40.6563 kJ mol^{-1}. Calculate ΔS for the transformation $H_2O(s,\ 0\ °C) \rightarrow H_2O(g,\ 100\ °C)$.

a) $$H_2O(\ell, 273.15\,\text{K}) \rightarrow H_2O(\ell, 373.15\,\text{K})$$

ΔS for the constant-pressure process is given by:

$$\Delta S = C_{p,m}\ \ln\frac{T_f}{T_i} = \left(75.291\ \text{J mol}^{-1}\ \text{K}^{-1}\right)\times\ln\frac{(373.15\ \text{K})}{(273.15\ \text{K})} = \underline{23.49\ \text{J mol}^{-1}\ \text{K}^{-1}}$$

b) For the process:

$$H_2O\,(s, 273.15\ \text{K}) \rightarrow H_2O\,(g, 373.15\ \text{K})$$

the following path has to be considered:

$$H_2O\,(s, 273.15\ \text{K}) \rightarrow H_2O\,(\ell, 273.15\ \text{K}) \rightarrow H_2O\,(\ell, 373.15\ \text{K}) \rightarrow H_2O\,(g, 373.15\ \text{K})$$

Therefore, ΔS_{fus} and ΔS_{vap} have to be calculated:

$$\Delta S_{fus} = \frac{\Delta H_{fus}}{T} = \frac{\left(6.0095\ \text{kJ mol}^{-1}\right)}{(273.15\ \text{K})} = \underline{22.00\ \text{J K}^{-1}}$$

$$\Delta S_{vap} = \frac{\Delta H_{fus}}{T} = \frac{\left(40.6563\ \text{kJ mol}^{-1}\right)}{(373.15\ \text{K})} = \underline{108.95\ \text{J K}^{-1}}$$

Finally, ΔS_{fus} for the entire process comes out to be:

$$\Delta S_{sys} = \Delta S_{fus} + C_{p,m}\left(\ell, H_2O\right)\ln\frac{T_f}{T_i} + \Delta S_{vap}$$

$$= \left(22.00\ \text{J K}^{-1}\right) + \left(75.3\ \text{J K}^{-1}\right)\times\ln\frac{(373.15\ \text{K})}{(273.15\ \text{K})} + \left(108.95\ \text{J K}^{-1}\right) = \underline{154.4\ \text{J K}^{-1}}$$

P5.15) Between 0.00 °C and 100. °C, the heat capacity of Hg(l) is given by

$$\frac{C_{P,m}\left(\text{Hg}, l\right)}{\text{J K}^{-1}\ \text{mol}^{-1}} = 30.093 - 4.944\times10^{-3}\,\frac{T}{\text{K}}$$

Calculate ΔH and ΔS if 1 mol of Hg(l) is raised in temperature from 0.00 ° to 100. °C at constant P.

With $C_{p,m} = A1 - A2\,T$, ΔH for the process is given by:

$$\Delta H = n\int_{T_1}^{T_2} C_{p,m}(T')\ dT' = n\int_{T_1}^{T_2} A1 - A2\,T'\,dT' = \left[A1\,T'\right]_{T_1}^{T_2} - \left[\tfrac{1}{2}A2\,T'^2\right]_{T_1}^{T_2} = A1\left(T_2 - T_1\right) - \tfrac{1}{2}A2\left(T_2^{\ 2} - T_1^{\ 2}\right)$$

$$= (1.0\ \text{mol})\times\left\{\begin{matrix}\left(30.093\ \text{J mol}^{-1}\ \text{K}^{-1}\right)\times\left(373.15\ \text{K} - 273.15\ \text{K}\right) \\ -\tfrac{1}{2}\times\left(4.944\times10^{-3}\ \text{J mol}^{-1}\ \text{K}^{-2}\right)\times\left(373.15^2\ \text{K}^2 - 273.15^2\ \text{K}^2\right)\end{matrix}\right\}$$

$= \underline{2.85\ \text{kJ}}$

ΔS for the process is given by:

$$\Delta S = n\int_{T_1}^{T_2}\frac{C_{p,m}(T')}{T'}\,dT' = n\int_{T_1}^{T_2}\frac{A1}{T'} - A2\,dT' = [A1\ln T']_{T_1}^{T_2} - [A2\,T']_{T_1}^{T_2} = A1\ln\frac{T_2}{T_1} - A2(T_2 - T_1)$$

$$= (1.0\text{ mol})\times\left\{\begin{aligned}&(30.093\text{ J mol}^{-1}\text{ K}^{-1})\times\ln\frac{(373.15\text{ K})}{(273.15\text{ K})}\\&-(4.944\times10^{-3}\text{ J mol}^{-1}\text{ K}^{-2})\times(373.15\text{ K} - 273.15\text{ K})\end{aligned}\right\}$$

$$= \underline{8.90\text{ J K}^{-1}}$$

P5.17) The heat capacity of α-quartz is given by

$$\frac{C_{P,m}(\alpha\text{-quartz}, s)}{\text{J K}^{-1}\text{ mol}^{-1}} = 46.94 + 34.31\times10^{-3}\frac{T}{\text{K}} - 11.30\times10^{5}\frac{T^2}{\text{K}^2}$$

The coefficient of thermal expansion is given by $\beta = 0.3530\times10^{-4}\text{ K}^{-1}$ and $V_m = 22.6$ $\text{cm}^3\text{ mol}^{-1}$. Calculate ΔS_m for the transformation α-quartz (25.0°C, 1 atm) → α-quartz (225 °C, 1000. atm).

$$\alpha - \text{Quartz}(1\text{ atm}, 298.15\text{ K}) \rightarrow \alpha - \text{Quartz}(1000\text{ atm}, 498\text{ K})$$

$$\Delta S = n\int_{T_i}^{T_2}\frac{C_{p,m}(T')}{T'} - V\beta(p_f - p_i)$$

With $C_{p,m} = A1 + A2\,T - A3\,T^2$:

$$n\int_{T_1}^{T_2}\frac{C_{p,m}(T')}{T'}\,dT' = n\int_{T_1}^{T_2}\left(\frac{A1}{T'} + A2 + A3\,T'\right)dT' = n\left(A1\ln\frac{T_2}{T_1} + A2(T_2 - T_1) - \tfrac{1}{2}A3(T_2^{\,2} - T_1^{\,2})\right)$$

$$= (1.0\text{ mol})\times\left\{\begin{aligned}&(46.94\text{ J mol}^{-1}\text{ K}^{-1})\times\ln\frac{(498.15\text{ K})}{(298.15\text{ K})}\\&+(34.31\times10^{-3}\text{ J mol}^{-1}\text{ K}^{-2})\times(498.15\text{ K} - 298.15\text{ K})\\&-\tfrac{1}{2}\times(11.3\times10^{-5}\text{ J mol}^{-1}\text{ K}^{-1})\times(498.15^2\text{ K}^2 - 298.15^2\text{ K}^2)\end{aligned}\right\}$$

$$= 21.95\text{ J K}^{-1}$$

$$V\beta(p_f - p_i) = (2.26\times10^{-5}\text{ m}^3)\times(0.3530\times10^{-4}\text{K}^{-1})\times(101325\text{ Pa} - 1000\times101325\text{ Pa}) = 0.008075\text{ J K}^{-1}$$

Therefore:

$$\Delta S = 21.95\text{ J K}^{-1} - 0.008075\text{ J K}^{-1} = \underline{21.87\text{ J K}^{-1}}$$

P5.18) The amino acid glycine dimerizes to form the dipeptide glycylglycine according to the reaction

$$2\text{Glycine}(s) \rightarrow \text{Glycylglycine}(s) + H_2O(l)$$

Calculate ΔS, $\Delta S_{surroundings}$, and $\Delta S_{universe}$ at $T = 298$ K. Useful thermodynamic data are:

	Glycine	Glycylglycine	Water
$\Delta H_f^\circ\left(\text{kJ mol}^{-1}\right)$	–537.2	–746.0	–285.8
$S^\circ(\text{J K}^{-1}\text{ mol}^{-1})$	103.5	190.0	70.0

The standard entropy for this reaction is:

$$\Delta S^\circ_{reaction} = \sum_i \nu_i S^\circ_{f,i} = -2\times\left(103.5\text{ J K}^{-1}\text{ mol}^{-1}\right)+\left(190\text{ J K}^{-1}\text{ mol}^{-1}\right)+\left(70\text{ J K}^{-1}\text{ mol}^{-1}\right)\underline{= 53\text{ J K}^{-1}\text{ mol}^{-1}}$$

The standard enthalpy for this reaction is:

$$\Delta H^\circ_{reaction} = \sum_i \nu_i H^\circ_{f,i} = -2\times\left(-537.2\text{ kJ mol}^{-1}\right)+\left(-746\text{ kJ mol}^{-1}\right)+\left(-285.8\text{ kJ mol}^{-1}\right)\underline{= 42.6\text{ kJ K}^{-1}\text{ mo}}$$

And the entropies for the surroundings and universe:

$$\Delta S_{surroundings} = \frac{-dq}{T} = \frac{-\Delta H^\circ_{reaction}}{T} = \frac{\left(-42.6\text{ kJ mol}^{-1}\right)}{(298.0\text{ K})}\underline{= -142.95\text{ J mol}^{-1}\text{ K}^{-1}}$$

$$\Delta S_{universe} = \Delta S_{reaction} + \Delta S_{surroundings} = \left(53\text{ J K}^{-1}\text{ mol}^{-1}\right)+\left(-142.95\text{ J K}^{-1}\text{ mol}^{-1}\right)\underline{= -89.95\text{ J mol}^{-1}\text{ K}^{-1}}$$

P5.25) One mole of $H_2O(l)$ is compressed from a state described by $P = 1.00$ bar and $T = 298$ K to a state described by $P = 800.$ bar and $T = 450.$ K. In addition, $\beta = 2.07 \times 10^{-4}\text{ K}^{-1}$ and the density can be assumed to be constant at the value 997 kg m^{-3}. Calculate ΔS for this transformation, assuming that $\kappa = 0$.

$$\Delta S = n\int_{T_1}^{T_2}\frac{C_{p,m}(T')}{T'} - V\beta(p_f - p_i)$$

$$n\int_{T_{low}}^{T_{high}} \frac{C_{p,m}(T')}{T'}\,dT' = n\int_{T_{low}}^{T_{high}}\left(\frac{A1}{T'}+A2+A3\,T'\right)dT' =$$

$$n\left(A1\ln\frac{T_{high}}{T_{low}}+A2\left(T_{high}-T_{low}\right)+\tfrac{1}{2}A3\left(T_{high}{}^{2}-T_{low}{}^{2}\right)+\tfrac{1}{3}A4\left(T_{high}{}^{3}-T_{low}{}^{3}\right)\right)$$

$$=(1.0\text{ mol})\times\left\{\begin{array}{l}\left(33.80\text{ J mol}^{-1}\text{ K}^{-1}\right)\times\ln\dfrac{(450.0\text{ K})}{(298.0\text{ K})}\\ +\left(-0.00795\text{ J mol}^{-1}\text{ K}^{-2}\right)\times(450.0\text{ K - }298.0\text{ K})\\ +\tfrac{1}{2}\times\left(2.8228\times10^{-5}\text{ J mol}^{-1}\text{ K}^{-3}\right)\times\left(450.0^{2}\text{ K}^{2}\text{ - }298.0^{2}\text{ K}^{2}\right)\\ +\tfrac{1}{3}\times\left(-1.3115\times10^{-8}\text{ J mol}^{-1}\text{ K}^{-4}\right)\times\left(450.0^{3}\text{ K}^{3}\text{ - }298.0^{3}\text{ K}^{3}\right)\end{array}\right\}$$

$$=14.04\text{ J K}^{-1}$$

$$V_m\beta(p_f-p_i)=\frac{\left(0.01802\text{ kg mol}^{-1}\right)\times\left(1\text{ m}^{3}\right)}{(997\text{ kg})}\times\left(2.07\times10^{-4}\text{ K}^{-1}\right)\times\left(800\times10^{5}\text{ Pa - }1\times10^{5}\text{ Pa}\right)$$

$$=0.2989\text{ J K}^{-1}\text{ mol}^{-1}$$

Therefore:

$$\Delta S=14.04\text{ J K}^{-1}\text{ mol}^{-1}+0.2989\text{ J K}^{-1}\text{ mol}^{-1}=\underline{14.34\text{ J K}^{-1}\text{ mol}^{-1}}$$

P5.28) The maximum theoretical efficiency of an internal combustion engine is achieved in a reversible Carnot cycle. Assume that the engine is operating in the Otto cycle and that $C_{V,m} = 5/2\,R$ for the fuel–air mixture initially at 298 K (the temperature of the cold reservoir). The mixture is compressed by a factor of 8.0 in the adiabatic compression step. What is the maximum theoretical efficiency of this engine? How much would the efficiency increase if the compression ratio could be increased to 30? Do you see a problem in doing so?

The temperatures and volumes are related by:

$$T_bV_c^{\gamma-1}=T_aV_d^{\gamma-1}$$

Solving for T_a yields for a compression ration of 1/8:

$$T_a=\frac{T_b\,V_c^{\gamma-1}}{V_d^{\gamma-1}}=T_b\left(\frac{V_c}{V_d}\right)^{\gamma-1}=(298\text{ K})\times\left(\frac{V_c}{8\times V_c}\right)^{-0.4}=(298\text{ K})\times\left(\frac{1}{8}\right)^{-0.4}-684.62\text{ K}$$

The efficiency of the cycle is then:

$$\varepsilon=1-\frac{T_{cold}}{T_{hot}}=1-\frac{(298\text{ K})}{(684.62\text{ K})}=\underline{0.565}$$

For a compression of 1/30 one obtains in analogy:

$$T_a = (298\text{ K})\times\left(\frac{1}{30}\right)^{-0.4} = 1161.623\text{ K}$$

$$\varepsilon = 1-\frac{T_{cold}}{T_{hot}} = 1-\frac{(298\text{ K})}{(1161.623\text{ K})} = \underline{0.743}$$

If the compression ration is too large, the fuel will ignite before the end of the compression cycle.

P5.29) One mole of $H_2O(l)$ is supercooled to –2.25 °C at 1 bar pressure. The freezing temperature of water at this pressure is 0.00 °C. The transformation $H_2O(l) \rightarrow H_2O(s)$ is suddenly observed to occur. By calculating ΔS, $\Delta S_{surroundings}$, and ΔS_{total}, verify that this transformation is spontaneous at –2.25 °C. The heat capacities are given by $C_P(H_2O(l))$ = 75.3 J K^{-1} mol^{-1} and $C_P(H_2O(s))$ = 37.7 J K^{-1} mol^{-1}, and ΔH_{fusion} = 6.008 kJ mol^{-1} at 0.00 °C. Assume that the surroundings are at –2.25°C. [*Hint:* Consider the two pathways at 1 bar: (a) $H_2O(l$, –2.25 °C) → $H_2O(s$, –2.25 °C) and (b) $H_2O(l$, –2.25 °C) →$H_2O(l$, 0.00 °C) → $H_2O(s$, 0.00 °C) → $H_2O(s$, –2.25 °C). Because S is a state function, ΔS must be the same for both pathways.]

ΔS can be calculated by considering the following path:

$$H_2O(\ell,270.9\text{ K}) \rightarrow H_2O(\ell,273.15\text{ K}) \rightarrow H_2O(s,273.15\text{ K}) \rightarrow H_2O(s,270.9\text{ K})$$

ΔS is given by:

$$\Delta S = C_{p,m}(\ell)\ln\frac{T_f}{T_i} - \frac{\Delta H_{fus}}{T} + C_{p,m}(s)\ln\frac{T_f}{T_i} = (1\text{ mol})\times(75.3\text{ J mol}^{-1}\text{ K}^{-1})\times\ln\frac{(273.15\text{ K})}{(270.9\text{ K})}$$
$$-\frac{(6.0095\text{ kJ mol}^{-1})}{(273.15\text{ K})} + (1\text{ mol})\times(37.7\text{ J mol}^{-1}\text{ K}^{-1})\times\ln\frac{(270.9\text{ K})}{(273.15\text{ K})} = \underline{-21.7\text{ J K}^{-1}}$$

$\Delta S_{sourroundings}$ is given by:

$$\Delta S_{surroundings} = \frac{\Delta H_{fus}}{T} = \frac{(6.0095\text{ kJ mol}^{-1})}{(270.9\text{ K})} = \underline{22.2\text{ J K}^{-1}}$$

And ΔS_{total} is then:

$$\Delta S_{total} = \Delta S_{surroundings} + \Delta S = (22.2\text{ J K}^{-1}) + (-21.7\text{ J K}^{-1}) = \underline{0.5\text{ J K}^{-1} > 0}$$

P5.31) The mean solar flux at the Earth's surface is ~4.00 J cm^{-2} min^{-1}. In a nonfocusing solar collector, the temperature can reach a value of 90.0°C. A heat engine is operated using the collector as the hot reservoir and a cold reservoir at 298 K. Calculate the area of the collector needed to produce one horsepower (1 hp = 746 watts). Assume that the engine operates at the maximum Carnot efficiency.

To obtain the area of the collector we use:

$$\frac{w_{cyc}}{q_{ab}} = 1 - \frac{T_{cold}}{T_{hot}}$$

w_{cyc} is set equal to 746 W (1 hp), and the amount of heat necessary to produce 746 W is:

$$q_{ab} = \frac{w_{cyc}}{\left(1 - \frac{T_{cold}}{T_{hot}}\right)} = \frac{\left(746\ \text{J s}^{-1}\right)}{\left(1 - \frac{(298\ \text{K})}{(363.15\ \text{K})}\right)} = 4158.25\ \text{J s}^{-1}$$

The collector produces the heat flux:

$$q_{collector} = 4.00\ \text{J min}^{-1} \times \frac{(1\ \text{min})}{(60\ \text{s})} = 0.067\ \text{J s}^{-1}$$

Therefore, the area for the collector needed is:

$$A_{collector} = \frac{\left(1\ \text{cm}^2\right) \times \left(4158.25\ \text{J s}^{-1}\right)}{\left(0.067\ \text{J s}^{-1}\right)} = \underline{62370\ \text{cm}^2 = 6.24\ \text{m}^2}$$

P5.34) Calculate ΔS, ΔS_{total}, and $\Delta S_{surroundings}$ when the volume of 85.0 g of CO initially at 298 K and 1.00 bar increases by a factor of three in (a) an adiabatic reversible expansion, (b) an expansion against $P_{external} = 0$, and (c) an isothermal reversible expansion. Take $C_{P,m}$ to be constant at the value 29.14 J mol^{-1}K^{-1} and assume ideal gas behavior. State whether each process is spontaneous. The temperature of the surroundings is 298 K.

a) For a reversible process with q = 0:

$$\Delta S = \Delta S_{sourroundings} = \Delta S_{tot} = 0\text{, not spontaneous}$$

b) For an expansion against $p_{ext} = 0$:

$$\Delta S = n\ R\ \ln\frac{V_f}{V_i} = \frac{(85\ \text{g})}{\left(28\ \text{g mol}^{-1}\right)} \times \left(8.314472\ \text{J mol}^{-1}\ \text{K}^{-1}\right) \times \ln\frac{(3 \times V_i)}{(V_i)} = \underline{27.7\ \text{J K}^{-1}}$$

$$\Delta S_{sorroundings} = \underline{0\ \text{J K}^{-1}}$$

$$\Delta S_{total} = \Delta S + \Delta S_{sorroundings} = \underline{27.7\ \text{J K}^{-1}}\text{, spontaneous}$$

c) For a reversible, isothermal expansion:

$$\Delta S = n\,R\ln\frac{V_f}{V_i} = \frac{(85\,g)}{(28\,g\,mol^{-1})}\times(8.314472\,J\,mol^{-1}\,K^{-1})\times\ln\frac{(3\times V_i)}{(V_i)} = \underline{27.7\,J\,K^{-1}}$$

$$\Delta S_{sorroundings} = -\Delta S = \underline{-27.7\,J\,K^{-1}}$$

$$\Delta S_{total} = \Delta S + \Delta S_{sorroundings} = \underline{0\,J\,K^{-1}}, \text{ not spontaneous}$$

P5.38) The interior of a refrigerator is typically held at 277 K and the interior of a freezer is typically held at 255 K. If the room temperature is 294 K, by what factor is it more expensive to extract the same amount of heat from the freezer than from the refrigerator? Assume that the theoretical limit for the performance of a reversible refrigerator is valid.

$$\eta(\text{freezer}) = \frac{T_{cold}}{(T_{hot} - T_{cold})} = \frac{(255\,K)}{(294\,K - 255\,K)} = 6.54$$

$$\eta(\text{fridge}) = \frac{T_{cold}}{(T_{hot} - T_{cold})} = \frac{(277\,K)}{(294\,K - 277\,K)} = 16.29$$

$$\frac{w_{freezer}}{w_{fridge}} = \frac{\eta_{fridge}}{\eta_{freezer}} = \underline{2.50}$$

P5.40) Calculate $\Delta S°$ for the reaction $H_2(g) + Cl_2(g) \rightarrow 2HCl(g)$ at 650. K. Omit terms in the temperature-dependent heat capacities higher than T^2/K^2.

$\Delta S°_{T=650\,K}$ for the reaction is given by:

$$\Delta S°_{T=650\,K} = \Delta S°_{T=298.15\,K} + \int_{T=298.15K}^{T=650K} \frac{n\,\Delta C°_{p,m}(T')}{T'}dT'$$

First:

$$\Delta S°_{T=298.15\,K} = \sum_i \nu_i\, S°_{f,i} =$$

$$(-130.7\,J\,K^{-1}\,mol^{-1}) + (-223.1\,J\,K^{-1}\,mol^{-1}) + 2\times(186.9\,J\,K^{-1}\,mol^{-1}) = 20.0\,J\,K^{-1}\,mol^{-1}$$

Then:

$$\int_{T=298.15K}^{T=650K} \frac{n\,\Delta C°_{p,m}(T')}{T'}dT' = \sum_i \int_{T=298.15K}^{T=650K} \frac{n\,\Delta C°_{p,m,i}(T')}{T'}dT' =$$

$$\sum_i n\int_{T_1}^{T_2}\left(\frac{A1_i}{T'} + A2_i + A3_i\,T'\right)dT' = \sum_i n\,A1_i\ln\left(\frac{T_2}{T_1}\right) + n\,A2_i(T_2 - T_1) + \tfrac{1}{2}n\,A3_i(T_2{}^2 - T_1{}^2)$$

$$\int_{T=298.15K}^{T=650K} \frac{n\,\Delta C^{\circ}_{p,m}(T')}{T'}dT' =$$

$$+(1\,\text{mol})\times\left\{\begin{array}{l}\left(22.66\ \text{J mol}^{-1}\ \text{K}^{-1}\right)\times\ln\dfrac{(650\ \text{K})}{(298.15\ \text{K})}\\ +\left(0.04381\ \text{J mol}^{-1}\ \text{K}^{-2}\right)\times(650\ \text{K}-298.15\ \text{K})\\ +\frac{1}{2}\times\left(-1.0835\times10^{-4}\ \text{J mol}^{-1}\ \text{K}^{-1}\right)\times\left(650^{2}\ \text{K}^{2}-298.15^{2}\ \text{K}^{2}\right)\end{array}\right\}$$

$$+(1\,\text{mol})\times\left\{\begin{array}{l}\left(22.85\ \text{J mol}^{-1}\ \text{K}^{-1}\right)\times\ln\dfrac{(650\ \text{K})}{(298.15\ \text{K})}\\ +\left(0.06543\ \text{J mol}^{-1}\ \text{K}^{-2}\right)\times(650\ \text{K}-298.15\ \text{K})\\ +\frac{1}{2}\times\left(-1.2517\times10^{-4}\ \text{J mol}^{-1}\ \text{K}^{-1}\right)\times\left(650^{2}\ \text{K}^{2}-298.15^{2}\ \text{K}^{2}\right)\end{array}\right\}$$

$$(2\,\text{mol})\times\left\{\begin{array}{l}\left(29.81\ \text{J mol}^{-1}\ \text{K}^{-1}\right)\times\ln\dfrac{(650\ \text{K})}{(298.15\ \text{K})}\\ +\left(-0.00412\ \text{J mol}^{-1}\ \text{K}^{-2}\right)\times(650\ \text{K}-298.15\ \text{K})\\ +\frac{1}{2}\times\left(6.2231\times10^{-4}\ \text{J mol}^{-1}\ \text{K}^{-1}\right)\times\left(650^{2}\ \text{K}^{2}-298.15^{2}\ \text{K}^{2}\right)\end{array}\right\}$$

$$=80.6\ \text{J K}^{-1}$$

And finally:

$$\Delta S^{\circ}_{T=650\,K} = 80.6\ \text{J K}^{-1} + 20.00\ \text{J K}^{-1} = \underline{100.6\ \text{J K}^{-1}}$$

P5.43) The following heat capacity data have been reported for L-alanine:

T (K)	10	20	40	60	80	100	140	180	220	260	300
$C^{\circ}_{P,m}\left(\text{J K}^{-1}\ \text{mol}^{-1}\right)$	0.49	3.85	17.45	30.99	42.59	52.50	68.93	83.14	96.14	109.6	122.7

By a graphical treatment, obtain the molar entropy of L-alanine at $T = 300$ K. [*Hint:* You can perform the integration numerically using either a spreadsheet program or a curve-fitting routine and a graphing calculator (see Example Problem 5.10).]

Plotting the data for L-alanine as $C^{0}_{p,m}$ versus T gives:

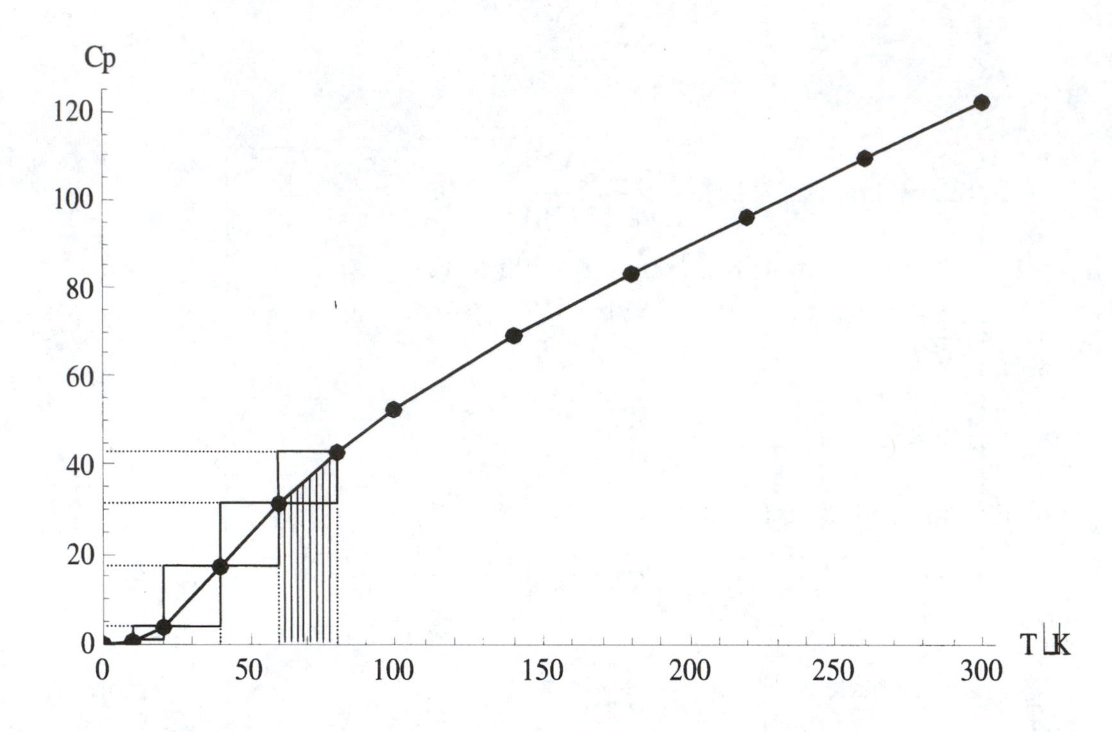

To obtain the molar entropy of L-alanine at 300 K we calculate the area under the curve for each temperature increment, divide the area by the upper temperature of the increment, and adding that number for all temperature increments. The area for each temperature increment can be obtained by calculating the areas under the red and blue step functions, adding the two, and dividing by two. For example the molar entropy for the increment (lined area):

$$S_m(60\,\mathrm{K}-40\,\mathrm{K})=\frac{(60\,\mathrm{K}-40\,\mathrm{K})\times(42.59\,\mathrm{J\,mol^{-1}\,K^{-1}})+(60\,\mathrm{K}-40\mathrm{K}\,)\times(30.99\,\mathrm{J\,mol^{-1}\,K^{-1}})}{2\times(60\,\mathrm{K})}$$

$$S_m(60\,\mathrm{K}-40\,\mathrm{K})=12.26\,\mathrm{J\,mol^{-1}\,K^{-1}}$$

Doing this calculation for all increments and adding yields:

$$S_m(\mathrm{L-alanine, 300\,K})\ \underline{=115.3\,\mathrm{J\,mol^{-1}\,K^{-1}}}$$

Chapter 6: The Gibbs Energy and Chemical Equilibrium

P6.1) Calculate the maximum nonexpansion work that can be gained from the combustion of benzene(*l*) and of $H_2(g)$ on a per gram and a per mole basis under standard conditions. Is it apparent from this calculation why fuel cells based on H_2 oxidation are under development for mobile applications?

At p = const., the nonexpansion work is:

$$\Delta G^{\circ}_{reaction} \leq dw_{nonexpansion}$$

So, for the combustion of benzene:

$$C_6H_6(\ell) + 7\tfrac{1}{2}O_2(g) \rightarrow 3CO_2(g) + 3H_2O(g)$$

$$dw_{nonexpansion} = \Delta G^{\circ}_{reaction} = \sum_i v_i G^{\circ}_{f,i} =$$

$$-1\times(124.5\text{ kJ mol}^{-1}) + 6\times(-394.4\text{ kJ mol}^{-1}) + 3\times(-228.6\text{ kJ mol}^{-1}) = -3176.7\text{ kJ mol}^{-1}$$

With the molecular mass of 78.12 g mol^{-1} of benzene, the nonexpansion work per gram of benzene is:

$$dw_{nonexpansion} = -3176.7\text{ kJ mol}^{-1} \times \frac{(1\text{ mol})}{(78.12\text{ g})} = \underline{-40.66\text{ kJ g}^{-1}}$$

Now in analogy, for the combustion of benzene:

$$H_2(g) + \tfrac{1}{2}O_2(g) \rightarrow H_2O(g)$$

$$dw_{nonexpansion} = \Delta G^{\circ}_{reaction} = \sum_i v_i G^{\circ}_{f,i} = -228.6\text{ kJ mol}^{-1}$$

With the molecular mass of 2.02 g mol^{-1} of H_2, the nonexpansion work per gram of H_2 is:

$$dw_{nonexpansion} = -228.6\text{ kJ mol}^{-1} \times \frac{(1\text{ mol})}{(2.02\text{ g})} = \underline{-113.17\text{ kJ g}^{-1}}$$

More than twice the work is produced per gram of H_2.

P6.3) Calculate ΔG for the isothermal expansion of 2.50 mol of an ideal gas at 350 K from an initial pressure of 10.5 bar to a final pressure of 0.500 bar.

ΔG for an isothermal expansion is given by:

$$\Delta G = n\,R\,T \ln\frac{p_f}{p_i} = (2.50\ \text{mol})\times(8.314472\ \text{J K}^{-1}\ \text{mol}^{-1})\times(350\ \text{K})\times\ln\frac{(10.5\times10^5\ \text{Pa})}{(0.5\times10^5\ \text{Pa})} = \underline{-22.15\ \text{J mol}^{-1}}$$

P6.5) The pressure dependence of G is quite different for gases and condensed phases. Calculate G_m(C, *solid, graphite,* 100 bar, 298.15 K) and G_m(He, *g,* 100 bar, 298.15 K) relative to their standard state values. By what factor is the change in G_m greater for He than for graphite?

For a liquid and solid:

$$\Delta G = \int_{p_i}^{p_f} V dp = V(p_f - p_i)$$

$$G_m(C,s,100\,bar) = G_m(C,s,1\,bar) + V_m(p_f - p_i) = G_m(C,s,1\,bar) + \frac{M}{\rho}(p_f - p_i)$$

$$G_m(\text{C,s,100 bar}) = (0) + \frac{(12.011\times10^{-3}\ \text{kg})}{(2250\ \text{kg m}^{-3})}\times(99.0\times10^5\,\text{Pa}) = 52.8\ \text{J}$$

Treating He as an ideal gas:

$$G_m(\text{He,g,100 bar}) = G_m(\text{He,g,1 bar}) + \int_{p_i}^{p_f} V\,dp = 0 + R\,T\ln\left(\frac{p_f}{p_i}\right)$$

$$G_m(\text{He,g,100 bar}) = (0) + (1\ \text{mol})\times(8.314472\ \text{J mol}^{-1}\ \text{K}^{-1})\times(298\ \text{K})\times\ln\left(\frac{(100\ \text{bar})}{(1\ \text{bar})}\right) = 11.4\times10^3\ \text{J}$$

Thus, the change in Gm is 216 times greater for He.

P6.8) Calculate $\Delta A^\circ_{reaction}$ and $\Delta G^\circ_{reaction}$ for the reaction $CH_4(g) + 2O_2(g) \rightarrow CO_2(g) + 2H_2O(l)$ at 298 K from the combustion enthalpy of methane and the entropies of the reactants and products:

$\Delta G^\circ_{\text{reaction}} = \Delta G^\circ_{\text{combustion}}$ is given by:

$$\Delta G^\circ_{\text{combustion}} = \Delta H^\circ_{\text{combustion}} - T\,\Delta S^\circ_{\text{combustion}}$$

$$\Delta S^\circ_{\text{combustion}} = \sum_i \nu_i\, S^\circ_{f,i} =$$

$$-1\times(186.3\ \text{J K}^{-1}\,\text{mol}^{-1}) - 2\times(205.2\ \text{J K}^{-1}\,\text{mol}^{-1}) + (213.8\ \text{J K}^{-1}\,\text{mol}^{-1}) + 2\times(70.0\ \text{J K}^{-1}\,\text{mol}^{-1})$$

$$= -\,242.9\ \text{J K}^{-1}\ \text{mol}^{-1}$$

From the table:

$\Delta H^\circ_{combustion}(CH_4) = -891.0\,\text{kJ mol}^{-1}$

Therefore:

$\Delta G^\circ_{combustion}(CH_4) = -891.0\,\text{kJ mol}^{-1} - (298.0\,\text{K}) \times \left(-242.9\,\text{J K}^{-1}\,\text{mol}^{-1}\right) = \underline{-818.6\,\text{kJ mol}^{-1}}$

$\Delta A^\circ_{reaction} = \Delta A^\circ_{combustion}$ is given by:

$$\Delta A^\circ_{combustion} = \Delta U^\circ_{combustion} - T\,\Delta S^\circ_{combustion} = \Delta H^\circ_{combustion} - \Delta n\,R\,T - T\,\Delta S^\circ_{combustion}$$
$$= \Delta G^\circ_{combustion} - \Delta n\,R\,T$$

$$\Delta A^\circ_{combustion} = -818.6\,\text{kJ mol}^{-1} - 2 \times \left(8.314472\,\text{J mol}^{-1}\,\text{K}^{-1}\right) \times (298.0\,\text{K}) = \underline{-813.6\,\text{kJ mol}^{-1}}$$

P6.14) Consider the equilibrium $NO_2\,(g) \rightleftarrows NO\,(g) + \frac{1}{2}\,O_2\,(g)$. One mole of $NO_2(g)$ is placed in a vessel and allowed to come to equilibrium at a total pressure of 1 bar. An analysis of the contents of the vessel gives the following results:

T	**700 K**	**800 K**
P_{NO}/P_{NO_2}	0.872	2.50

a. Calculate K_P at 700 and 800 K.

b. Calculate $\Delta G^\circ_{reaction}$ for this reaction at 298.15 K, assuming that $\Delta H^\circ_{reaction}$ is independent of temperature.

a) K_p for the reaction is given by:

$$K_p = \frac{\left(\frac{p(NO)}{p^\circ}\right)^1 \times \left(\frac{p(O_2)}{p^\circ}\right)^{1/2}}{\left(\frac{p(NO_2)}{p^\circ}\right)^1} = \frac{p(NO)}{p(NO_2)} \times \left(\frac{p(O_2)}{p^\circ}\right)^{1/2} = \frac{(x_{NO} \times p)}{(x_{NO_2} \times p)} \times \frac{(x_{O_2} \times p)^{1/2}}{p^{\circ 1/2}}$$

The mol fractions for the gases are calculated as follows:

$$\frac{p_{NO}}{p_{NO_2}} = \frac{x_{NO}}{x_{NO_2}} = \frac{n_{NO}}{n_{NO_2}}$$

Therefore, at 700 K:

$x_{NO} = n_{NO} = 0.872 \times n_{NO_2}$

Also:

$x_{O_2} = \frac{1}{2} \times 0.872 \times x_{NO_2}$

$$1 = x_{NO} + x_{NO_2} + x_{O_2} = 0.872 \times x_{NO_2} + x_{NO_2} + \tfrac{1}{2} \times 0.872 \times x_{NO_2} = x_{NO_2} \times \left(0.872 + 1 + \tfrac{1}{2} \times 0.872\right)$$

$$x_{NO_2} = 0.433$$

$$x_{NO} = 0.872 \times x_{NO_2} = 0.872 \times 0.433 = 0.378$$

$$x_{O_2} = \tfrac{1}{2} \times 0.872 \times x_{NO_2} = \tfrac{1}{2} \times 0.872 \times 0.433 = 0.189$$

So, Kp at 700 K is:

$$K_p(700\,K) = \frac{(x_{NO} \times p)}{(x_{NO_2} \times p)} \times \frac{(x_{O_2} \times p)^{1/2}}{p^{\circ 1/2}} \frac{(0.378 \times 1\,bar)}{(0.433 \times 1 bar)} \times \frac{(0.189 \times 1\,bar)^{1/2}}{(1\,bar)^{1/2}} \underline{= 0.379}$$

At 800 K:

$$1 = x_{NO} + x_{NO_2} + x_{O_2} = 2.50 \times x_{NO_2} + x_{NO_2} + \tfrac{1}{2} \times 2.50 \times x_{NO_2} = x_{NO_2} \times \left(2.50 + 1 + \tfrac{1}{2} \times 2.50\right)$$

$$x_{NO_2} = 0.211$$

$$x_{NO} = 2.50 \times x_{NO_2} = 2.50 \times 0.211 = 0.528$$

$$x_{O_2} = \tfrac{1}{2} \times 2.50 \times x_{NO_2} = \tfrac{1}{2} \times 2.50 \times 0.211 = 0.264$$

So Kp at 800 K is:

$$K_p(800\,K) = \frac{(x_{NO} \times p)}{(x_{NO_2} \times p)} \times \frac{(x_{O_2} \times p)^{1/2}}{p^{\circ 1/2}} \frac{(0.528 \times 1\,bar)}{(0.211 \times 1 bar)} \times \frac{(0.264 \times 1\,bar)^{1/2}}{(1\,bar)^{1/2}} \underline{= 1.286}$$

b) ΔG can be obtained from K_p:

$$\Delta G_{reaction}(700\,K) = -\ln K_p\, R\, T = -\ln(0.379) \times \left(8.314472\,J\,mol^{-1}\,K^{-1}\right) \times (700.0\,K) = 5646.8\,J\,mol^{-1}$$

ΔH is assumed to be independent of T:

$$\Delta H_{reaction} = \sum_i \nu_i\, H^{\circ}_{f,i} = \left(-33.2\,kJ\,mol^{-1}\right) + \left(91.3\,kJ\,mol^{-1}\right) = 58.1\,kJ\,mol^{-1}$$

Then $\Delta G_{reaction}(298.15\,K)$ can be calculated using:

$$\Delta G_{reaction}(T_2) = T_2 \times \left[\frac{\Delta G_{reaction}(T_1)}{T_1} + \Delta H_{reaction}\left(\frac{1}{T_2} - \frac{1}{T_1}\right)\right]$$

$$\Delta G_{reaction}(298.15\,K) = (298.15\,K) \times \left[\frac{\left(5646.8\,J\,mol^{-1}\right)}{(700.0\,K)} + \left(58.1\,kJ\,mol^{-1}\right)\left(\frac{1}{(298.15\,K)} - \frac{1}{(700.0\,K)}\right)\right]$$

$$\underline{= 35.8\,kJ\,mol^{-1}}$$

P6.22) For a protein denaturation the entropy change is 2.31 J K^{-1} mol^{-1} at P =1.00 bar and at the melting temperature T = 338 K. Calculate the melting temperature at a pressure of $P = 1.00 \times 10^3$ bar if the heat capacity change $\Delta C_{P,m}$ = 7.98 J $K^{-1}mol^{-1}$ and if ΔV = 3.10 mL mol^{-1}. State any assumptions you make in the calculation.

At 1.00 bar we have:

$$\Delta H_m^0(T_m) = T_m\,\Delta S_m^0 = (338\text{ K})\times(2.31\text{ kJ mol}^{-1}\text{ K}^{-1}) = 780.8\text{ kJ mol}^{-1}$$

At 1000 bar: As pressure increases the thermodynamic potentials will change, changing T_m as well. This is described by the Clapeyron equation:

$$\frac{dp}{dT} = \frac{\Delta H_m^0}{T\,\Delta V_m}$$

H describes the slope of the line at equilibrium in a p/T phased diagram and therefore leads us to the new melting (denaturating) temperature, T_m, at 1000 bar after we integrate.

We assume that ΔV_m and $\Delta C_{p,m}$ stay constant.

$$\Delta C_{p,m} = \left(\frac{\partial \Delta H_m^0}{\partial T}\right)$$

$$\Delta H_m^0(T) = \Delta H_m^0(T_m) + \int_{T_m}^{T} \Delta C_{p,m}\,dT = \Delta H_m^0(T_m) + \int_{T_m}^{T} \Delta C_{p,m}(T - T_m)$$

We plug this back into the Clapeyron equation and obtain:

$$\int_{p_i}^{p_f} dp = \int_{T_m}^{T_n} \frac{1}{\Delta V_m}\left\{\Delta H_m^0(T_m) + \Delta C_{p,m}T - \Delta C_{p,m}T_m\right\}\frac{1}{T}\,dT$$

$$\Delta p = \frac{1}{\Delta V_m}\left\{\left[\Delta H_m^0(T_m) - \Delta C_{p,m}T_m\right]\ln\left(\frac{T_n}{T_m}\right) + \Delta C_{p,m}(T_n - T_m)\right\}$$

$$(100\times10^5\text{ Pa}) - (1\times10^5\text{ Pa}) = \frac{1}{(0.0000031\text{ m}^3)}$$

$$\times\left\{\left[(780.8\text{ kJ mol}^{-1}) - (7.98\text{ J mol}^{-1}\text{ K}^{-1})\times(338\text{ K})\right]\ln\left(\frac{T_n}{T_m}\right) + (7.98\text{ J mol}^{-1}\text{ K}^{-1})\times\Delta T\right\}$$

$$(99\times10^5\text{ N m}^{-2}) = (2.51\times10^{11}\text{ N m}^{-2})\times\left(\frac{T_n}{T_m}\right) + (2.57\times10^6\text{ N m}^{-2})\times\Delta T$$

The second term can be neglected relative to the first term and we obtain:

$$\ln\left(\frac{T_n}{(338\ \text{K})}\right) = \left(3.98\times10^{-4}\right)$$

$$\underline{T_n = 338.1\ \text{K}}$$

P6.28) Calculate $\mu_{O_2}^{mixture}(298.15\ \text{K},\ 1\ \text{bar})$ for oxygen in air, assuming that the mole fraction of O_2 in air is 0.200.

For pure O_2:

$$\mu_{O_2}(\text{pure}) = \Delta G_m^\circ(O_2) = -T\,\Delta S_m^\circ(O_2) = -(298.15\ \text{K})\times\left(205.2\ \text{J mol}^{-1}\ \text{K}^{-1}\right) = -61180.38\ \text{J mol}^{-1}$$

For O_2 in air:

$$\mu_{O_2}(\text{mix}) = \mu_{O_2}(\text{pure}) + R\,T\ln x_{O_2}$$

$$= \left(-61180.38\ \text{J mol}^{-1}\right) + \left(8.314472\ \text{J K}^{-1}\ \text{mol}^{-1}\right)\times(298.15\ \text{K})\times\ln(0.200) = \underline{-65.2\ \text{kJ mol}^{-1}}$$

P6.33) Calculate K_P at 550. K for the reaction $N_2O_4(l) \rightarrow 2NO_2(g)$ assuming that $\Delta H_{reaction}^\circ$ is constant over the interval from 298 to 600 K.

To get Kp for the reaction at 550 K, ΔG at 550 K has to be determined, while assuming that ΔH is independent of T over the temperature range considered, and we can use ΔH at 298.15 K:

$$\Delta G_{\text{reaction}}^\circ(298.15K) = \sum_i v_i\, G_{f,i}^\circ = -1\times\left(97.5\ \text{kJ mol}^{-1}\right) + 2\times\left(51.3\ \text{kJ mol}^{-1}\right) = 5.1\ \text{kJ mol}^{-1}$$

$$\Delta H_{\text{reaction}}^\circ(298.15K) = \sum_i v_i\, H_{f,i}^\circ = -1\times\left(-19.5\ \text{kJ mol}^{-1}\right) + 2\times\left(33.2\ \text{kJ mol}^{-1}\right) = 85.9\ \text{kJ mol}^{-1}$$

$$\Delta G_{\text{reaction}}(T_2) = T_2\times\left[\frac{\Delta G_{\text{reaction}}(T_1)}{T_1} + \Delta H_{\text{reaction}}\left(\frac{1}{T_2} - \frac{1}{T_1}\right)\right]$$

$$\Delta G_{\text{reaction}}(550\ \text{K}) = (550\ \text{K})\times\left[\frac{\left(5.1\ \text{kJ mol}^{-1}\right)}{(298.15\ \text{K})} + \left(85.9\ \text{kJ mol}^{-1}\right)\left(\frac{1}{(550\ \text{K})} - \frac{1}{(298015\ \text{K})}\right)\right]$$

$$\underline{= -63152.65\ \text{J mol}^{-1}}$$

Then:

$$K_p(550\text{ K}) = \text{Exp}\left[-\frac{\Delta G(550\text{ K})}{R\,T}\right] = \text{Exp}\left[-\frac{(-63152.65\text{ J mol}^{-1})}{(8.314472\text{ J mol}^{-1}\text{ K}^{-1})\times(550\text{ K})}\right] = 9.95\times10^{5}$$

P6.35) Calculate the degree of dissociation of N_2O_4 in the reaction $N_2O_4(g) \rightarrow 2NO_2(g)$ at 250 K and a total pressure of 0.500 bar. Do you expect the degree of dissociation to increase or decrease as the temperature is increased to 550 K? Assume that $\Delta H^{\circ}_{reaction}$ is independent of temperature.

We set up the table:

	N_2O_4	$2\,NO_2$
Initial number of moles	n_0	0
Moles at equilibrium	$n_0 - \zeta$	2ζ
Mole fractions at equilibrium	$\frac{n_0 - \zeta}{n_0 + \zeta}$	$\frac{2\zeta}{n_0 + \zeta}$
Partial pressures at equilibrium, $p_i = x_i\,p$	$\left(\frac{n_0 - \zeta}{n_0 + \zeta}\right)p$	$\left(\frac{2\zeta}{n_0 + \zeta}\right)p$

We next express K_p in terms of n_0, ζ, and p:

$$K_p(T) = \frac{\left(\frac{p^{eq}_{NO_2}}{p^0}\right)^2}{\left(\frac{p^{eq}_{N_2O_4}}{p^0}\right)} = \frac{\left[\left(\frac{2\zeta}{n_0 + \zeta}\right)\left(\frac{p}{p^0}\right)\right]^2}{\left(\frac{n_0 - \zeta}{n_0 + \zeta}\right)\left(\frac{p}{p^0}\right)} = \frac{4\zeta^2}{(n_0 + \zeta)(n_0 - \zeta)}\frac{p}{p^0} = \frac{4\zeta^2}{n_0^{\,2} - \zeta^2}\frac{p}{p^0}$$

We convert this expression for K_p to one in terms of α:

$$K_p(T) = \frac{4\zeta^2}{n_0^{\,2} - \zeta^2}\frac{p}{p^0} = \frac{4\alpha^2}{n_0^{\,2} - \alpha^2}\frac{p}{p^0}$$

And solving for K_p yields:

$$\alpha = \sqrt{\frac{K_p(T)}{\left(K_p(T) + 4\frac{p}{p^0}\right)}}$$

ΔG and ΔH of the reaction are given by:

$$\Delta G^{\circ}_{reaction}(298.15K) = \sum_i \nu_i \, G^{\circ}_{f,i} = -1\times(97.5\,\text{kJ mol}^{-1}) + 2\times(51.3\,\text{kJ mol}^{-1}) = 5.1\,\text{kJ mol}^{-1}$$

$$\Delta H^{\circ}_{reaction}(298.15K) = \sum_i \nu_i \, H^{\circ}_{f,i} = -1\times(-19.5\,\text{kJ mol}^{-1}) + 2\times(33.2\,\text{kJ mol}^{-1}) = 85.9\,\text{kJ mol}^{-1}$$

$$\ln(K_p)(T_f) = -\frac{\Delta G(298.15\,K)}{R\,T_i} - \frac{\Delta H_{reaction}}{R}\left(\frac{1}{T_f} - \frac{1}{T_i}\right)$$

$$\ln(K_p)(250\,K) = -\frac{(5100\,\text{J mol}^{-1})}{(8.314472\,\text{J mol}^{-1}\,\text{K}^{-1})\times(298.15\,\text{K})} - \frac{(85900\,\text{J mol}^{-1})}{(8.314472\,\text{J mol}^{-1}\,\text{K}^{-1})}\left(\frac{1}{(250\,\text{K})} - \frac{1}{(298.15\,\text{K})}\right)$$

$$\ln(K_p)(250\,K) = -5.426$$

$$K_p(250\,K) = 4.40\times10^{-3}$$

And:

$$\alpha = \sqrt{\frac{(4.40\times10^{-3})}{((4.40\times10^{-3}) + 4\times0.5)}} = 4.68\times10^{-2}$$

Because $\Delta H^{0}_{reaction} > 0$, α increases as T increases.

P6.37) A sample containing 2.00 mol of N_2 and 6.00 mol of H_2 is placed in a reaction vessel and brought to equilibrium at 20.0 bar and 750 K in the reaction $1/2N_2(g) + 3/2H_2(g) \rightarrow NH_3(g)$.

a. Calculate K_P at this temperature.

b. Set up an equation relating K_P and the extent of reaction as in Example Problem 6.10.

c. Using a numerical equation solver, calculate the number of moles of each species present at equilibrium.

a) ΔG and ΔH of the reaction are given by:

$$\Delta G^{\circ}_{reaction}(298.15\,K) = -16.5\,\text{kJ mol}^{-1}$$

$$\Delta H^{\circ}_{reaction}(298.15\,K) = -45.9\,\text{kJ mol}^{-1}$$

K_p at 298.15 K is then:

$$K_p(298.15\,K) = \text{Exp}\left[-\frac{\Delta G(298.15\,K)}{R\,T}\right] = \text{Exp}\left[-\frac{(-16500\,\text{J mol}^{-1})}{(8.314472\,\text{J mol}^{-1}\,\text{K}^{-1})\times(298.15\,\text{K})}\right] = 777.45$$

And K_p at 750 K:

$$K_p(750\,\text{K}) = \text{Exp}\left[\ln(777.45) - \frac{(-45900\,\text{J mol}^{-1})}{(8.314472\,\text{J mol}^{-1}\,\text{K}^{-1})}\left(\frac{1}{(750\,\text{K})} - \frac{1}{(298.15\,\text{K})}\right)\right] = \underline{1.11\times10^{-2}}$$

b) The degree of dissociation at 750 K:

$$\alpha(750\,\text{K}) = \sqrt{\frac{K_p(750\,\text{K})}{\left(K_p(750\,\text{K}) + 4\frac{p}{p^\circ}\right)}} = \sqrt{\frac{1.11\times10^{-2}}{\left(1.11\times10^{-2} + 4\times\frac{20\,\text{bar}}{1.0\,\text{bar}}\right)}} = \underline{0.0118}$$

c) Setting up a table with initial and equilibrium number of moles, and the equilibrium mol fractions, where ξ denotes the number of moles of NH_3 at equilibrium:

	N_2	H_2	NH_3
n_{ini}	2	6	0
n_{eq}	$2-\frac{1}{2}\xi$	$6-\frac{3}{2}\xi$	ξ
$x_{i,eq}$	$\frac{(2-\frac{1}{2}\xi)}{(8-\xi)}$	$\frac{(6-\frac{3}{2}\xi)}{(8-\xi)}$	$\frac{\xi}{(8-\xi)}$

The total number of moles at equilibrium is calculated as:

$$n_{tot,eq} = 2-\tfrac{1}{2}\xi + 6 - \tfrac{3}{2}\xi + \xi = 8-\xi$$

$$K_p = \frac{\left(\frac{p(NH_3)}{p^\circ}\right)^1 \times}{\left(\frac{p(N_2)}{p^\circ}\right)^{1/2}\left(\frac{p(H_2)}{p^\circ}\right)^{3/2}} = \frac{p(NH_3)}{p(N_2)^{1/2}p(H_2)^{3/2}}\times p^\circ = \frac{(x_{NH_3}\times p)}{(x_{N_2}\times p)^{1/2}(x_{H_2}\times p)^{3/2}}\times p^\circ = 0.0111$$

$$\frac{\left(\frac{\xi}{(8-\xi)}\times(20\,\text{bar})\right)}{\left(\frac{(2-\frac{1}{2}\xi)}{(8-\xi)}\times(20\,\text{bar})\right)^{1/2}\left(\frac{(6-\frac{3}{2}\xi)}{(8-\xi)}\times(20\,\text{bar})\right)^{3/2}}\times(1\,\text{bar}) = 0.0111$$

Using some appropriate software to solve the equation for the equilibrium number of moles of NH_3, ξ, yields:

$$\underline{n(NH_3)_{eq} = \xi = 0.47}$$

And then:

$$\underline{n(N_2)_{eq} = 2-\tfrac{1}{2}\xi = 2-\tfrac{1}{2}\times 0.47 = 1.76}$$

$$\underline{n(H_2)_{eq} = 6-\tfrac{3}{2}\xi = 6-\tfrac{3}{2}\times 0.47 = 5.30}$$

P6.38) Consider the equilibrium in the reaction $3\,O_2\,(g) \rightleftarrows 2\,O_3\,(g)$ with $\Delta H^\circ_{reaction} = 285.4\times 10^3$ J mol^{-1} at 298 K. Assume that $\Delta H^\circ_{reaction}$ is independent of temperature.

- **a.** Without doing a calculation, predict whether the equilibrium position will shift toward reactants or products as the pressure is increased.
- **b.** Without doing a calculation, predict whether the equilibrium position will shift toward reactants or products as the temperature is increased.
- **c.** Calculate K_P at 550 K.
- **d.** Calculate K_x at 550 K and 0.500 bar.

a) Since the number of moles on the product side (2) is smaller than on the reactant side (3), increasing the pressure would shift the equilibrium towards the product side.

b) Since the reaction is endothermic ($\Delta H > 0$), increasing the temperature would shift the equilibrium towards the products.

c) To obtain K_p at 550 K, first K_p at 298.15 has to be calculated. This requires the calculation of $\Delta S^\circ_{reaction}$, and subsequently $\Delta G^\circ_{reaction}$ at 298.15 K:

$$\Delta S^\circ_{reaction}(298.15\,K) = \sum_i \nu_i\, S^\circ_{f,i} = 2\times(238.9\,J\,K^{-1}\,mol^{-1}) - 3\times(205.2\,J\,K^{-1}\,mol^{-1}) = -137.8\,J\,K^{-1}\,mol^{-1}$$

$$K_p(298.15\,K) = \mathrm{Exp}\left[-\frac{\Delta G(298.15\,K)}{R\,T}\right]$$

$$= \mathrm{Exp}\left[-\frac{\{(285.4\times 10^3\,J\,mol^{-1}) + (298.15K)\times(-137.8\,J\,K^{-1}\,mol^{-1})\}}{(8.314472\,J\,mol^{-1}\,K^{-1})\times(550\,K)}\right] = 5.9865\times 10^{-58}$$

Then:

$$K_p(550\,K) = \mathrm{Exp}\left[\ln(298.15\,K) - \frac{(285.4\times 10^3\,J\,mol^{-1})}{(8.314472\,J\,mol^{-1}\,K^{-1})}\left(\frac{1}{(550\,K)} - \frac{1}{(298.15\,K)}\right)\right] = \underline{4.98\times 10^{-35}}$$

d) K_x at 550 K and 0.500 bar with $\Delta\nu = 1$ is given by:

$$K_x = K_p\left(\frac{p}{p^\circ}\right)^{-\Delta\nu} = K_p\left(\frac{0.5\,bar}{1.0\,bar}\right)^{-1} = 2\,K_p = 2\times(4.98\times 10^{-35}) = \underline{9.96\times 10^{-35}}$$

P6.44) The standard Gibbs energy for the formation of double-stranded DNA from single strands follows the equation $\Delta G^\circ = \Delta G^\circ_{initiation} + \sum \Delta G^\circ_{neighbors}$. The Gibbs energy of initiation for two terminal A—T pairs is $\Delta G^\circ_{initiation} \approx 8.10 \text{ kJ mol}^{-1}$ when two separated strands come together to form a duplex. $\Delta G^\circ_{neighbors}$ is the contribution to the Gibbs energy from the hydrogen bonding and stacking as successive nucleotide pairs are added. As such, this contribution is characteristic of the nearest neighbor pairs. A table of nearest-neighbor Gibbs energies for DNA duplex formation at T = 310.K is given as follows:

	5′-A-A-3′ 3′-T-T-5′	**5′-A-T-3′ 3′-T-A-5′**	**5′-A-G-3′ 3′-T-C-5′**	**5′-G-C-3′ 3′-C-G-5′**	**5′-C-G-3′ 3′-G-C-5′**	**5′-C-A-3′ 3′-G-T-5′**	**5′-T-A-3′ 3′-A-T-5′**
$\Delta G^\circ_{neighbors}$ $\left(\text{kJmol}^{-1}, 310\text{K}\right)$	–4.2	–3.7	–5.4	–9.3	–9.1	–6.0	–2.4
$\Delta H^\circ_{neighbors}$ $\left(\text{kJmol}^{-1}\right)$	–33.1	–30.2	–32.7	–41.0	–44.4	–35.6	–30.2
$\Delta S^\circ_{neighbors}$ $\left(\text{JK}^{-1}\text{ mol}^{-1}\right)$	–93.0	–85.4	–87.9	– 102	–114	–95.0	–89.2

Source: H. T. Allawai & J. SantaLucia, Jr. 1997, *Biochemistry*, 36, 10581–10594.

Consider the duplex formation reaction, where the dot indicates a Watson-Crick base pair:

$$\begin{matrix} 5'-A-G-C-G-C-A-3' \\ + \\ 3'-T-C-G-C-G-T-5' \end{matrix} \rightleftharpoons \begin{matrix} 5'-A-G-C-G-C-A-3' \\ \cdot\ \cdot\ \cdot\ \cdot\ \cdot\ \cdot \\ 3'-T-C-G-C-G-T-5' \end{matrix}$$

Calculate the Gibbs energy change and equilibrium constant for this reaction at T = 310.K.

The Gibbs energy change for the duplex formation is:

$$\Delta G_{formation} = \Delta G^0_{initiation} + \sum \Delta G^0_{neighbors}$$

$$\Delta G_{formation} = \left(8.1 \text{ kJ mol}^{-1}\right) + \left(-5.4 \text{ kJ mol}^{-1}\right) + \left(-9.3 \text{ kJ mol}^{-1}\right)$$
$$+ \left(-9.1 \text{ kJ mol}^{-1}\right)\left(-9.3 \text{ kJ mol}^{-1}\right) + \left(-6.0 \text{ kJ mol}^{-1}\right) = \underline{-31.0 \text{ kJ mol}^{-1}}$$

The equilibrium constant is:

$$K = \text{Exp}\left[\frac{-\Delta G_{\text{formation}}}{R\,T}\right] = \text{Exp}\left[\frac{-\left(-31.0\times10^{3}\ \text{J mol}^{-1}\right)}{\left(8.314472\ \text{J mol}^{-1}\ \text{K}^{-1}\right)\times\left(310\ \text{K}\right)}\right] = \underline{1.67\times10^{5}}$$

Chapter 7: Phase Equilibria

P7.3) Within what range can you restrict the values of P and T if the following information is known about CO_2? Use Figure 7.8 to answer this problem.

a. As the temperature is increased, the solid is first converted to the liquid and subsequently to the gaseous state.

b. As the pressure on a cylinder containing pure CO_2 is increased from 65 to 80 atm, no interface delineating liquid and gaseous phases is observed.

c. Solid, liquid, and gas phases coexist at equilibrium.

d. An increase in pressure from 10 to 50 atm converts the liquid to the solid.

e. An increase in temperature from –80 °C to 20 °C converts a solid to a gas with no intermediate liquid phase.

Using Figure 7.8

a) $-56.6\ ^\circ C \le T \le 31.0\ ^\circ C$

$5.11\ atm \le p \le 73.75\ atm$

b) $T > T_c > 31.0\ ^\circ C$

c) $T = -56.6\ ^\circ C$ and $p = 5.11$ atm

d) $1\ atm \le p \le 5.11\ atm$

P7.14) It has been suggested that the surface melting of ice plays a role in enabling speed skaters to achieve peak performance. Carry out the following calculation to test this hypothesis. At 1 atm pressure, ice melts at 273.15 K, $\Delta H_{fusion} = 6010\ J\ mol^{-1}$, the density of ice is 920 kg m^{-3}, and the density of liquid water is 997 kg m^{-3}.

a. What pressure is required to lower the melting temperature by 5.0 °C?

b. Assume that the width of the skate in contact with the ice has been reduced by sharpening to 25×10^{-3} cm, and that the length of the contact area is 15 cm. If a skater of mass 85 kg is balanced on one skate, what pressure is exerted at the interface of the skate and the ice?

c. What is the melting point of ice under this pressure?

d. If the temperature of the ice is –5.0 °C, do you expect melting of the ice at the ice-skate interface to occur?

a) The pressure that would result in a 5 °C lower melting temperature can be obtained using:

$$p_f - p_i = \frac{\Delta H^{fus}}{\Delta V_m^{fus}} \ln\left(\frac{T_f}{T_i}\right)$$

To calculate ΔV_m^{fus} the densities of ice and water are used:

$$\Delta V_m^{fus}(\text{ice}) = \frac{M\,n}{d} = \frac{\left(0.01802\ \text{kg mol}^{-1}\right)\times\left(1.0\ \text{mol}\right)}{\left(920\ \text{kg m}^{-3}\right)} = 1.9587\times10^{-5}\ \text{m}^{-3}$$

$$\Delta V_m^{fus}(\text{water}) = \frac{M\,n}{d} = \frac{\left(0.01802\ \text{kg mol}^{-1}\right)\times\left(1.0\ \text{mol}\right)}{\left(997\ \text{kg m}^{-3}\right)} = 1.8074\times10^{-5}\ \text{m}^{-3}$$

Solving for p_f yields:

$$p_f = \frac{\Delta H^{fus}}{\Delta V_m^{fus}} \ln\left(\frac{T_f}{T_i}\right) + p_i$$

$$= \frac{\left(6010\ \text{J mol}^{-1}\right)}{\left\{\left(1.8074\times10^{-5}\ \text{mol}^{-1}\ \text{m}^{-3}\right) - \left(1.9587\times10^{-5}\ \text{mol}^{-1}\ \text{m}^{-3}\right)\right\}} \times \ln\left(\frac{268.15\ \text{K}}{273.15\ \text{K}}\right) + 101325\ \text{Pa}$$

$$\underline{= 7.3427\times10^{7}\ \text{Pa} = 734.3\ \text{bar}}$$

b) The pressure exerted by the skater is given by:

$$p = \frac{F}{A} = \frac{m\,g}{A} = \frac{\left(9.80665\ \text{m s}^{-2}\right)\times\left(85\ \text{kg}\right)}{\left(0.15\ \text{m}\right)\times\left(2.5\times10^{-4}\ \text{m}\right)} \underline{= 2.22\times10^{7}\ \text{kg m}^{-1}\ \text{s}^{-2} = 2.22\times10^{7}\ \text{Pa} = 2.22\times10^{2}\ \text{bar}}$$

c) To obtain the melting temperature at the pressure exerted by the skater we use:

$$p_f - p_i = \frac{\Delta H^{fus}}{\Delta V_m^{fus}} \ln\left(\frac{T_f}{T_i}\right)$$

And solve for T_f:

$$T_f = \text{Exp}\left[\frac{\left(p_f - p_i\right)}{\left(\frac{\Delta H^{fus}}{\Delta V_m^{fus}}\right)} + \ln T_i\right] = \text{Exp}\left[\frac{\left\{\left(2.22\times10^{7}\ \text{Pa}\right) - \left(101325\ \text{Pa}\right)\right\}}{\left(\frac{\left(6010\ \text{J mol}^{-1}\right)}{\left(-3.9728\times10^{-6}\ \text{mol}^{-1}\ \text{m}^{-3}\right)}\right)} + \ln\left(273.15\ \text{K}\right)\right]$$

$$\underline{= 271.62\ \text{K} = -1.53\ ^{\circ}\text{C}}$$

d) No, the temperature of the ice is below the melting temperature resulting from the pressure exerted by the skater.

P7.15) Solid iodine, $I_2(s)$, at 25 °C has an enthalpy of sublimation of 56.30 kJ mol^{-1}. The $C_{P,m}$ of the vapor and solid phases at that temperature are 36.9 and 54.4 J K^{-1} mol^{-1}, respectively. The sublimation pressure at 25 °C is 0.30844 Torr. Calculate the sublimation pressure of the solid at the melting point (113.6 °C) assuming

a. that the enthalpy of sublimation is constant.

b. that the enthalpy of sublimation at temperature T can be calculated from the equation

$$\Delta H^{\circ}_{sublimation}(T) = \Delta H^{\circ}_{sublimation}(T_0) + \Delta C_P (T - T_0).$$

a) The pressure of sublimation at 113.6 °C can be calculated using:

$$\ln\left(\frac{p_f}{p_i}\right) = -\frac{\Delta H^{fus}}{R}\left(\frac{1}{T_f} - \frac{1}{T_i}\right)$$

Solving for p_f yields:

$$p_f = \text{Exp}\left[-\frac{\Delta H^{fus}}{R}\left(\frac{1}{T_f} - \frac{1}{T_i}\right) + \ln(p_f)\right]$$

$$= \text{Exp}\left[-\frac{(56.30\text{ kJ mol}^{-1})}{(8.314472\text{ J mol}^{-1}\text{ K}^{-1})} \times \left(\frac{1}{(386.75\text{ K})} - \frac{1}{(298.15\text{ K})}\right) + \ln(0.30844\text{ Torr})\right] = \underline{56.07\text{ Torr}}$$

P7.17) Consider the transition between two forms of solid tin, Sn(*s*, gray) ⇄ Sn(*s*, white). The two phases are in equilibrium at 1 bar and 18 °C. The densities for gray and white tin are 5750 and 7280 kg m^{-3}, respectively, and $\Delta H_{transition}$ = 8.8 J K^{-1} mol^{-1}. Calculate the temperature at which the two phases are in equilibrium at 200 bar.

We use the Clapeyron equation:

$$(p_f - p_i) = \frac{\Delta H}{\Delta V_m}\left(\frac{(T_f - T_i)}{T_1}\right)$$

and solve for T_i:

$$T_f = \frac{(V_m(w) - V_m(g))T_i(p_f - p_i)}{\Delta H}$$

$$= \frac{\left(\left(\frac{(118.7\times10^{-3}\text{ kg mol}^{-1})}{(7280\text{ kg m}^3)}\right) - \left(\frac{(118.7\times10^{-3}\text{ kg mol}^{-1})}{(5750\text{ kg m}^3)}\right)\right)\times(291.5\text{ K})\times(200\times10^5\text{ Pa} - 1\times10^5\text{ Pa})}{(8.8\text{ kJ mol}^{-1})}$$

$= \underline{15.1\ ^{\circ}\text{C}}$

P7.19) A protein has a melting temperature of T_m = 335 K. At T = 315 K, UV absorbance determines that the fraction of native protein is f_N = 0.965. At T = 345. K, f_N = 0.015. Assuming a two-state model and assuming also that the enthalpy is constant between T = 315 and 345 K, determine the enthalpy of denaturation. Also, determine the entropy of denaturation at T = 335 K. By DSC, the enthalpy of denaturation was determined to be 251 kJ mol^{-1}. Is this denaturation accurately described by the two-state model?

a) To determine ΔH^{den}, the equilibrium constants at 315 K and 345 K have to be determined:

$$K(315\,K)=\frac{f_D}{f_N}=\frac{1-f_N}{f_N}\frac{(1-0.965)}{0.965}=0.03627$$

$$K(345\,K)=\frac{f_D}{f_N}=\frac{1-f_N}{f_N}\frac{(1-0.015)}{0.015}=65.6667$$

Then:

$$\Delta H^{den}=\frac{-\ln\left(\frac{K_2}{K_1}\right)R}{\left(\frac{1}{T_2}-\frac{1}{T_1}\right)}=\frac{-\ln\left(\frac{65.6667}{0.03627}\right)(8.314472\text{ J mol}^{-1}\text{ K}^{-1})}{\left(\frac{1}{(345\text{ K})}-\frac{1}{(315\text{ K})}\right)}=\underline{225.9\text{ kJ mol}^{-1}}$$

b) At equilibrium, K = 1 and therefore:

$$\Delta S_m^{den}=\frac{\Delta H^{den}}{T_m}=\frac{225.9\text{ kJ mol}^{-1}}{335\text{ K}}=\underline{674.4\text{ J mol}^{-1}\text{ K}^{-1}}$$

P7.21) The vapor pressure of methanol(*l*) is given by

$$\ln\left(\frac{P}{\text{Pa}}\right)=23.593-\frac{3.6791\times10^3}{\frac{T}{\text{K}}-31.317}$$

a. Calculate the standard boiling temperature.

b. Calculate $\Delta H_{vaporization}$ at 298 K and at the standard boiling temperature.

a) Solving the equation given for the vapor pressure of methanol (l) as a function of temperature yields for the standard boiling temperature:

$$T(K)=\frac{3.6791\times10^3}{(-\ln(p/\text{Pa})+23.593)}+31.317=\frac{3.6791\times10^3}{(-\ln(101325\text{ Pa})+23.593)}+31.317=\underline{336.2\text{ K}}$$

b) ΔH_{vap} at 298 K and at the standard boiling temperature (336.2 K from a) can be estimated by calculating the vapor pressures of methanol in a small temperature range. For T = 298 K:

$$p(297.5\text{ K}) = \text{Exp}\left[23.593 - \frac{3.6791\times10^3}{(297.5-31.317)}\right] = 17523.6\text{ Pa}$$

$$p(298.5\text{ K}) = \text{Exp}\left[23.593 - \frac{3.6791\times10^3}{(298.5-31.317)}\right] = 18454.0\text{ Pa}$$

$$\Delta H^{fus} = \frac{-\ln\left(\frac{p_f}{p_i}\right)R}{\left(\frac{1}{T_f}-\frac{1}{T_i}\right)} = \frac{-\ln\left(\frac{18454.0\text{ Pa}}{17523.6\text{ Pa}}\right)(8.314472\text{ J mol}^{-1}\text{ K}^{-1})}{\left(\frac{1}{(298.5\text{ K})}-\frac{1}{(297.5\text{ K})}\right)} = \underline{38.20\text{ kJ mol}^{-1}}$$

For T = 336.2 K:

$$p(335.7\text{ K}) = \text{Exp}\left[23.593 - \frac{3.6791\times10^3}{(335.7-31.317)}\right] = 99302.4\text{ Pa}$$

$$p(336.7\text{ K}) = \text{Exp}\left[23.593 - \frac{3.6791\times10^3}{(336.7-31.317)}\right] = 103312.0\text{ Pa}$$

$$\Delta H^{fus} = \frac{-\ln\left(\frac{p_f}{p_i}\right)R}{\left(\frac{1}{T_f}-\frac{1}{T_i}\right)} = \frac{-\ln\left(\frac{103312.0\text{ Pa}}{99302.4\text{ Pa}}\right)(8.314472\text{ J mol}^{-1}\text{ K}^{-1})}{\left(\frac{1}{(336.7\text{ K})}-\frac{1}{(335.7\text{ K})}\right)} = \underline{37.20\text{ kJ mol}^{-1}}$$

P7.29) The densities of a given solid and liquid of molecular weight 122.5 at its normal melting temperature of 427.15 K are 1075 and 1012 kg m^{-3}, respectively. If the pressure is increased to 120 bar, the melting temperature increases to 429.35 K. Calculate $\Delta H^{\circ}_{fusion}$ and $\Delta S^{\circ}_{fusion}$ for this substance.

The mol volumes for the solid and liquid can be calculated as:

$$V_m^{fus}(\text{solid}) = \frac{M\,n}{d} = \frac{(0.1225\text{ kg mol}^{-1})\times(1.0\text{ mol})}{(1075\text{ kg m}^{-3})} = 1.1395\times10^{-4}\text{ m}^{-3}$$

$$V_m^{fus}(\text{liquid}) = \frac{M\,n}{d} = \frac{(0.1225\text{ kg mol}^{-1})\times(1.0\text{ mol})}{(1012\text{ kg m}^{-3})} = 1.2105\times10^{-4}\text{ m}^{-3}$$

Then:

$$\Delta V_m^{fus} = V_m^{fus}(\text{liquid}) - V_m^{fus}(\text{solid}) = \left(1.2105\times10^{-4}\ \text{m}^{-3}\right) - \left(1.1395\times10^{-4}\ \text{m}^{-3}\right) = 7.097\times10^{-6}\ \text{m}^{-3}$$

Solving:

$$p_f - p_i = \frac{\Delta H^{fus}}{\Delta V_m^{fus}}\frac{\Delta T}{T_i}$$

for ΔH^{fus} yields:

$$\Delta H^{fus} = \frac{(p_f - p_i)\Delta V_m^{fus}\,T_i}{\Delta T} = \frac{\left(120\times10^5\ \text{Pa}\right)\times\left(7.097\times10^{-6}\ \text{m}^{-3}\right)\times(427.15\ \text{K})}{(429.35\ \text{K} - 427.15\ \text{K})} = \underline{16.4\ \text{kJ mol}^{-1}}$$

$$\Delta S^{fus} = \frac{\Delta H^{fus}}{T_m} = \frac{\left(16.4\ \text{kJ mol}^{-1}\right)}{(427.15\ \text{K})} = \underline{38.4\ \text{J mol}^{-1}\ \text{K}^{-1}}$$

P7.31) Suppose the denaturation temperature of a protein is $T = 325$ K and $\Delta H° = 245$ kJ mol^{-1}. Assuming the denaturation of a protein can be described by a two state model, calculate the equilibrium constant and standard Gibbs energy change for the denaturation of the protein at $T = 310.$ K. Assume the enthalpy change is constant between $T = 310.$ K and $T = 325$ K.

At the denaturation, the equilibrium constant is K = 1. We can therefore use:

$$\ln\left(\frac{K_2}{K_1}\right) = -\frac{\Delta H^{\circ}_{den}}{R}\left(\frac{1}{T_2} - \frac{1}{T_1}\right),$$

with $K_1 = 1$ and $T_1 = T_m = 325$ K, and solve for K_2 at T = 310 K, to obtain:

$$K_2 = \text{Exp}\left[-\frac{\Delta H^{\circ}_{den}}{R}\left(\frac{1}{T_2} - \frac{1}{T_1}\right) + \ln(K_1)\right]$$

$$= \text{Exp}\left[-\frac{\left(245000\ \text{J mol}^{-1}\right)}{\left(8.314472\ \text{J mol}^{-1}\text{K}^{-1}\right)}\left(\frac{1}{(310\ \text{K})} - \frac{1}{(325\ \text{K})}\right) + \ln(1)\right]$$

$$\underline{K_2 = 0.0124}$$

ΔG°_{den} is then obtained with:

$$\Delta G^{\circ}_{den} = -R\,T\ln(K) = \left(8.314472\ \text{J mol}^{-1}\text{K}^{-1}\right)\times(310\ \text{K})\times\ln(0.0124) = \underline{11.3\ \text{kJ mol}^{-1}}$$

P7.32) The variation of the vapor pressure of the liquid and solid forms of a pure substance near the triple point are given by $\ln(P_{solid}/\text{Pa}) = -8750(\text{K}/T) + 31.143$ and $\ln(P_{liquid}/\text{Pa}) = -4053(\text{K}/T) + 21.10$. Calculate the temperature and pressure at the triple point.

At the triple point, the vapor pressures are equal, therefore:

$$\frac{(-8750\,\mathrm{K})}{\mathrm{T}}+31.143=\frac{(-4053\,\mathrm{K})}{\mathrm{T}}+21.10$$

Solving for T yields:

$\underline{\mathrm{T}=467.7\,\mathrm{K}}$

Then, the pressure at the triple point is:

$$\mathrm{p}=\mathrm{Exp}\left[-\frac{(8750\,\mathrm{K})}{(467.7\,\mathrm{K})}+31.143\right]=\underline{2.513\times10^{5}\,\mathrm{Pa}}$$

P7.39) Autoclaves that are used to sterilize surgical tools require a temperature of 120 °C to kill bacteria. If water is used for this purpose, at what pressure must the autoclave operate? The normal boiling point of water is 373.15 K, and $\Delta H^{\circ}_{vaporication}=40.656\times10^{3}\ \mathrm{J\ mol^{-1}}$ at the normal boiling point.

We can use:

$$\ln\left(\frac{\mathrm{p_f}}{\mathrm{p_i}}\right)=-\frac{\Delta\mathrm{H}^{fus}}{\mathrm{R}}\left(\frac{1}{\mathrm{T_f}}-\frac{1}{\mathrm{T_i}}\right),$$

and solve for p_f:

$$\mathrm{p_f}=\mathrm{Exp}\left[-\frac{\Delta\mathrm{H}^{fus}}{\mathrm{R}}\left(\frac{1}{\mathrm{T_f}}-\frac{1}{\mathrm{T_i}}\right)+\ln(\mathrm{p_f})\right]$$

$$=\mathrm{Exp}\left[-\frac{(40.656\,\mathrm{kJ\,mol^{-1}})}{(8.314472\,\mathrm{J\,mol^{-1}\,K^{-1}})}\times\left(\frac{1}{(393.15\,\mathrm{K})}-\frac{1}{(373.15\,\mathrm{K})}\right)+\ln(1)\right]=\underline{1.95\,\mathrm{atm}}$$

P7.41) Calculate the factor by which the vapor pressure of a droplet of methanol of radius 1.00×10^{-4} m at 45.0 °C in equilibrium with its vapor is increased with respect to a very large droplet. Use the tabulated value of the density and the surface tension at 298 K from Appendix B for this problem. (*Hint:* You need to calculate the vapor pressure of methanol at this temperature.)

We can use:

$$\ln\left(\frac{\mathrm{p}}{\mathrm{p^0}}\right)=\frac{2\gamma\mathrm{M}}{\mathrm{r}\rho\mathrm{R\,T}},$$

therefore:

$$\frac{p}{p^0} = \mathrm{Exp}\left[\frac{2\gamma M}{r\,\rho\,R\,T}\right] = \mathrm{Exp}\left[\frac{2\times\left(22.07\times10^{-3}\ \mathrm{N\,m^{-1}}\right)\times\left(0.03204\ \mathrm{kg\,mol^{-1}}\right)}{\left(1.00\times10^{-4}\ \mathrm{m}\right)\times\left(791.4\ \mathrm{kg\,m^{-3}}\right)\times\left(8.314472\ \mathrm{J\,mol^{-1}\,K^{-1}}\right)\times\left(318.15\ \mathrm{K}\right)}\right]$$

$$\underline{= 1.000067}$$

Chapter 8: Ideal and Real Solutions

P8.1) At 303 K, the vapor pressure of benzene is 118 Torr and that of cyclohexane is 122 Torr. Calculate the vapor pressure of a solution for which $x_{benzene} = 0.25$ assuming ideal behavior.

The vapor pressure of the solution is given by:

$$p_{tot} = x_1 p_1^* + (1 - x_1) p_1^* = 0.25 \times 118 \text{ Torr} + (1 - 0.25) \times 122 \text{ Torr} = \underline{121 \text{ Torr}}$$

P8.3) An ideal solution is formed by mixing liquids A and B at 298 K. The vapor pressure of pure A is 180. Torr and that of pure B is 82.1 Torr. If the mole fraction of A in the vapor is 0.450, what is the mole fraction of A in the solution?

The mol fraction of A in the ideal solution of liquids A and B is given by:

$$x_A = \frac{y_A p_B^*}{p_A^* + (p_B^* - p_A^*) y_A} = \frac{0.450 \times 82.1 \text{ Torr}}{180 \text{ Torr} + (82.1 \text{ Torr} - 180 \text{ Torr}) \times 0.450} = \underline{0.272}$$

P8.5) A and B form an ideal solution at 298 K, with $x_A = 0.600$, $P_A^* = 105 \text{ Torr}$, and $P_B^* = 63.5 \text{ Torr}$.

a. Calculate the partial pressures of A and B in the gas phase.

b. A portion of the gas phase is removed and condensed in a separate container. Calculate the partial pressures of A and B in equilibrium with this liquid sample at 298 K.

a) The gas phase pressures of A and B are:

$$p_A = x_A p_A^* = 0.600 \times 105 \text{ Torr} = \underline{63 \text{ Torr}}$$

$$p_B = x_B p_B^* = (1 - x_A) p_B^* = (1 - 0.600) \times 63.5 \text{ Torr} = \underline{25.4 \text{ Torr}}$$

P8.6) The vapor pressures of 1-bromobutane and 1-chlorobutane can be expressed in the form

$$\ln\frac{P_{bromo}}{\text{Pa}} = 17.076 - \frac{1584.8}{\frac{T}{\text{K}} - 111.88}$$

and

$$\ln\frac{P_{chloro}}{\text{Pa}} = 20.612 - \frac{2688.1}{\frac{T}{\text{K}} - 55.725}$$

Assuming ideal solution behavior, calculate x_{bromo} and y_{bromo} at 300.0 K and a total pressure of 8741 Pa.

We use:

$$x_{bromo} = \frac{y_{bromo}\, p^*_{chloro}}{p^*_{bromo} + \left(p^*_{chloro} - p^*_{bromo}\right)y_{bromo}}$$

$$y_{bromo} = \frac{p^*_{bromo}\, p_{tot} - p^*_{bromo}\, p^*_{chloro}}{p_{tot}\left(p^*_{bromo} - p^*_{chloro}\right)}$$

First, we need to calculate the vapor pressures for 1-bromobutane and 1-chlorobutane:

$$\ln\left(\frac{p_{vap}(\text{bromo})}{\text{Pa}}\right) = 17.076 - \frac{1584.8}{\frac{\text{T}}{\text{K}} - 111.88} = 17.076 - \frac{1584.8}{300.0 - 111.88} = 8.6516$$

$$p_{vap}(\text{bromo}) = 5719.2\ \text{Pa}$$

$$\ln\left(\frac{p_{vap}(\text{chloro})}{\text{Pa}}\right) = 20.612 - \frac{2688.1}{\frac{\text{T}}{\text{K}} - 55.725} = 20.612 - \frac{2688.1}{300 - 55.725} = 9.6076$$

$$p_{vap}(\text{chloro}) = 14877.4\ \text{Pa}$$

Then:

$$y_{chloro} = \frac{(5719.2\ \text{Pa})\times(8741\ \text{Pa}) - (5719.2\ \text{Pa})\times(14877.4\ \text{Pa})}{(8741\ \text{Pa})\times((5719.2\ \text{Pa}) - (14877.4\ \text{Pa}))} = \underline{0.44}$$

$$x_{bromo} = \frac{0.48\times(14877.4\ \text{Pa})}{(5719.2\ \text{Pa}) + ((14877.4\ \text{Pa}) - (5719.2\ \text{Pa}))\times 0.44} = \underline{0.67}$$

P8.8) An ideal solution at 298 K is made up of the volatile liquids A and B, for which $P_A^* = 125\ \text{Torr}$ and $P_B^* = 46.3\ \text{Torr}$. As the pressure is reduced from 450 Torr, the first vapor is observed at a pressure of 70.0 Torr. Calculate x_A.

70 Torr corresponds to the total pressure of the ideal mixture. Solving

$p_{tot} = x_A p_A^* + (1 - x_A) p_B^*$ for x_A yields:

$$x_A = \frac{(p_{tot} - p_B^*)}{(p_A^* - p_B^*)} = \frac{(70\ \text{Torr} - 46.3\ \text{Torr})}{(125\ \text{Torr} - 46.3\ \text{Torr})} = 0.301$$

P8.10) At –31.2 °C, pure propane and *n*-butane have vapor pressures of 1200 and 200 Torr, respectively.

a. Calculate the mole fraction of propane in the liquid mixture that boils at –31.2 °C at a pressure of 760 Torr.

b. Calculate the mole fraction of propane in the vapor that is in equilibrium with the liquid of part (a).

a) At the boiling point the vapor pressure of the mixture is equal to the external pressure:

$p_{tot} = p_{ext} = 760\ \text{Torr}$

Again, solving $p_{tot} = x_{prop} p_{prop}^* + (1 - x_{prop}) p_{but}^*$ for x_A yields for the mol fraction of propane in the solution:

$$x_{prop} = \frac{(p_{tot} - p_{but}^*)}{(p_{prop}^* - p_{but}^*)} = \frac{(760\ \text{Torr} - 200\ \text{Torr})}{(1200\ \text{Torr} - 200\ \text{Torr})} = 0.560$$

b) The mol fraction of propane in the gas phase is:

$$y_{prop} = \frac{p_{but}^* p_{tot} - p_{but}^* p_{prop}^*}{p_{tot}(p_{but}^* - p_{prop}^*)} = \frac{(200\ \text{Torr}) \times (760\ \text{Torr}) - (200\ \text{Torr}) \times (1200\ \text{Torr})}{(760\ \text{Torr}) \times ((200\ \text{Torr}) - (1200\ \text{Torr}))} = 0.884$$

P8.11) In an ideal solution of A and B, 3.50 mol are in the liquid phase and 4.75 mol are in the gaseous phase. The overall composition of the system is $Z_A = 0.300$ and $x_A = 0.250$. Calculate y_A.

Solving the lever rule:

$$n_{liq}^{tot}(Z_A - x_A) = n_{vap}^{tot}(y_B - Z_B)$$

for the vapor mol fraction of A, y_A yields:

$$y_B = \frac{\left(n_{liq}^{tot}(Z_A - x_A) + n_{vap}^{tot} Z_B\right)}{n_{vap}^{tot}} = \frac{(3.50\ \text{mol} \times (0.300 - 0.250) + 4.75\ \text{mol} \times 0.300)}{4.75\ \text{mol}} = \underline{0.337}$$

P8.13) At 39.9 °C, a solution of ethanol ($x_1 = 0.9006$, $P_1^* = 130.4$ Torr) and isooctane ($P_2^* = 43.9$ Torr) forms a vapor phase with $y_1 = 0.6667$ at a total pressure of 185.9 Torr.

a. Calculate the activity and activity coefficient of each component.

b. Calculate the total pressure that the solution would have if it were ideal.

a) For the non-ideal solution the pressure for ethanol, p_A, is given by:

$$p_A = y_A p_{tot} = 0.6667 \times (185.9\ \text{Torr}) = 123.94\ \text{Torr}$$

Then the activity and the activity coefficient for ethanol are given by:

$$a_A = \frac{p_{tot}}{p_A^*} = \frac{(123.94\ \text{Torr})}{(130.4\ \text{Torr})} = \underline{0.9504}$$

$$\gamma_A = \frac{a_A}{x_A} = \frac{0.9504}{0.9066} = \underline{1.055}$$

To get the activity and the activity coefficient for component B, isooctane, we get through the nonideal pressure, p_B:

$$p_B = p_{tot} - p_A = (185.9\ \text{Torr}) - (123.9\ \text{Torr}) = 61.96\ \text{Torr}$$

And then:

$$a_B = \frac{p_B}{p_B^*} = \frac{(61.96\ \text{Torr})}{(43.9\ \text{Torr})} = \underline{1.411}$$

$$\gamma_B = \frac{a_B}{x_B} = \frac{a_B}{(1 - x_A)} = \frac{1.411}{(1\text{-}0.9006)} = \underline{14.20}$$

b) The total pressure in an ideal would be:

$$p_{tot} = x_A p_A^* + (1 - x_A) p_B^* = 0.9006 \times (130.4\ \text{Torr}) + (1 - 0.9006) \times (43.9\ \text{Torr}) = \underline{121.8\ \text{Torr}}$$

P8.19) A solution is prepared by dissolving 32.5 g of a nonvolatile solute in 200. g of water. The vapor pressure above the solution is 21.85 Torr and the vapor pressure of pure water is 23.76 Torr at this temperature. What is the molecular weight of the solute?

To get the molecular weight of the solute we have to calculate the number of moles. Solving for n_{solute} yields:

$$\frac{p_{solvent}}{p^*_{solvent}} = 1 - \frac{n_{solute}}{(n_{solute} + n_{solvent})}$$

$$n_{solute} = \frac{\left(-n_{solvent} + n_{solvent}\frac{p_{solvent}}{p^*_{solvent}}\right)}{\left(\frac{p_{solvent}}{p^*_{solvent}}\right)} = \frac{\left(-\frac{m_{solvent}}{M_{solvent}} + \frac{m_{solvent}}{M_{solvent}}\frac{p_{solvent}}{p^*_{solvent}}\right)}{\left(\frac{p_{solvent}}{p^*_{solvent}}\right)}$$

$$n_{solute} = \frac{\left(-\frac{(200\,\mathrm{g})}{(18.02\,\mathrm{g\,mol^{-1}})} + \frac{(200\,\mathrm{g})}{(18.02\,\mathrm{g\,mol^{-1}})}\frac{(21.85\,\mathrm{Torr})}{(23.76\,\mathrm{Torr})}\right)}{\left(\frac{(21.85\,\mathrm{Torr})}{(23.76\,\mathrm{Torr})}\right)} = 0.9702\,\mathrm{mol}$$

Then:

$$M_{solute} = \frac{n_{solute}}{m_{solute}} = \frac{(32.5\,\mathrm{g})}{(0.9702\,\mathrm{mol})} = \underline{33.5\,\mathrm{g\,mol^{-1}}}$$

P8.21) The dissolution of 5.25 g of a substance in 565 g of benzene at 298 K raises the boiling point by 0.625 °C. Note that $K_f = 5.12$ K kg mol^{-1}, $K_b = 2.53$ K kg mol^{-1}, and the density of benzene is 876.6 kg m^{-3}. Calculate the freezing point depression, the ratio of the vapor pressure above the solution to that of the pure solvent, the osmotic pressure, and the molecular weight of the solute. Note that $P^*_{benzene} = 103$ Torr at 298 K.

a) To calculate the freezing point depression we need the molality, which can be obtained through the boiling point raising:

$\Delta T_b = k_b\, m_{solute}$, therefore: $m_{solute} = \dfrac{\Delta T_b}{k_b}$

$$\Delta T_f = -k_f\, m_{solute} = -k_f\frac{\Delta T_b}{k_b} = -(5.12\,\mathrm{K\ kg\,mol^{-1}}) \times \frac{(0.625\,\mathrm{K})}{(2.53\,\mathrm{K\ kg\,mol^{-1}})} = \underline{-1.265\,\mathrm{K}}$$

b) The ratio of pressures is given by:

$$m_{solute} = \frac{\Delta T_b}{k_b} = \frac{(0.625\,\mathrm{K})}{(2.53\,\mathrm{K\ kg\,mol^{-1}})} = 0.247\,\mathrm{mol\,kg^{-1}}$$

$$x_b = \frac{n_b}{n_{tot}} = \frac{n_b}{(n_b + n_{substance})} = \frac{\left(\frac{(565\text{ g})}{(78.12\text{ g mol}^{-1})}\right)}{\left(\frac{(565\text{ g})}{(78.12\text{ g mol}^{-1})}\right) + \left(\frac{(0.247\text{ mol kg}^{-1})\times(565\text{ g})}{(1000\text{ g kg}^{-1})}\right)} = \frac{(7.2325\text{ mol})}{(7.3720\text{ mol})} = 0.981$$

c) The osmotic pressure is given by:

$$\pi = \frac{n_{solute}RT}{V} = \frac{\left(\frac{(0.247\text{ mol kg}^{-1})\times(565\text{ g})}{(1000\text{ g kg}^{-1})}\right)\times(8.314472\text{ J mol}^{-1}\text{ K}^{-1})\times(298\text{ K})}{\left(\frac{(0.565\text{ kg})}{(876.6\text{ kg m}^3)}\right)} = 5.38\times10^5\text{ Pa}$$

d) The molecular mass of the substance is:

$$M_{substance} = \frac{m_{substance}}{n_{substance}} = \frac{(5.25\text{ g})}{(0.14\text{ mol})} = 37.5\text{ g mol}^{-1}$$

P8.29) Calculate the solubility of H_2S in 1 L of water if its pressure above the solution is 3.25 bar. The density of water at this temperature is 997 kg m^{-3}.

The solubility of H_2S in 1 L of H_2O is given by:

$$n_{H_2S} = n_{H_2O}\frac{p_{H_2S}}{k_H(H_2O)} = \left(\frac{(997\text{ kg m}^{-3})\times(0.001\text{ m}^3)}{(0.01802\text{ kg mol}^{-1})}\right)\times\frac{(3.25\text{ bar})}{(5.68\times10^2\text{ bar})} = 0.317\text{ mol}$$

P8.31) At a given temperature, a nonideal solution of the volatile components A and B has a vapor pressure of 832 Torr. For this solution, $y_A = 0.404$. In addition, $x_A = 0.285$, $P_A^* = 591\text{ Torr}$, and $P_B^* = 503$ Torr. Calculate the activity and activity coefficient of A and B.

a) For the nonideal solution the pressure for A is given by:

$$p_A = y_A p_{tot} = 0.404\times(832\text{ Torr}) = 336.13\text{ Torr}$$

Then the activity and the activity coefficient for A are given by:

$$a_A = \frac{p_{tot}}{p_A^*} = \frac{(336.13\text{ Torr})}{(591\text{ Torr})} = 0.569$$

$$\gamma_A = \frac{a_A}{x_A} = \frac{0.569}{0.285} = 2.00$$

To get the activity and the activity coefficient for component B we get through the nonideal pressure, p_B:

$$p_B = p_{tot} - p_A = (832\text{ Torr}) - 336.13 = 495.87\text{ Torr}$$

And then:

$$a_B = \frac{p_B}{p_B^*} = \frac{(495.87\text{ Torr})}{(503\text{ Torr})} = \underline{0.986}$$

$$\gamma_B = \frac{a_B}{x_B} = \frac{a_B}{(1 - x_A)} = \frac{0.986}{(1 - 0.285)} = \underline{1.38}$$

Chapter 9: Electrolyte Solutions, Electrochemical Cells, and Redox Reactions

P9.1) Calculate $\Delta H^{\circ}_{reaction}$ and $\Delta G^{\circ}_{reaction}$ for the reaction $AgNO_3(aq) + KCl(aq) \rightarrow AgCl(s) + KNO_3(aq)$.

$\Delta H^{\circ}_{reaction}$ and $\Delta G^{\circ}_{reaction}$ are given by:

$$\Delta H^{\circ}_{reaction} = \Delta H^{\circ}_{f}(AgCl(s)) + \Delta H^{\circ}_{f}(K^{+}(aq)) + \Delta H^{\circ}_{f}(NO_3^{-}(aq))$$
$$- \Delta H^{\circ}_{f}(K^{+}(aq)) - \Delta H^{\circ}_{f}(Cl^{-}(aq)) - \Delta H^{\circ}_{f}(Ag^{+}(aq)) - \Delta H^{\circ}_{f}(NO_3^{-}(aq))$$

$$\Delta H^{\circ}_{reaction} = (-127.0\text{ kJ mol}^{-1}) + (-252.4\text{ kJ mol}^{-1}) + (-207.4\text{ kJ mol}^{-1})$$
$$-(-252.4\text{ kJ mol}^{-1}) - (-167.2\text{ kJ mol}^{-1}) - (105.6\text{ kJ mol}^{-1}) - (-207.4\text{ kJ mol}^{-1})$$
$$\underline{= -65.4\text{ kJ mol}^{-1}}$$

$$\Delta G^{\circ}_{reaction} = \Delta G^{\circ}_{f}(AgCl(s)) + \Delta G^{\circ}_{f}(K^{+}(aq)) + \Delta G^{\circ}_{f}(NO_3^{-}(aq))$$
$$- \Delta G^{\circ}_{f}(K^{+}(aq)) - \Delta G^{\circ}_{f}(Cl^{-}(aq)) - \Delta G^{\circ}_{f}(Ag^{+}(aq)) - \Delta G^{\circ}_{f}(NO_3^{-}(aq))$$

$$\Delta G^{\circ}_{reaction} = (-109.8\text{ kJ mol}^{-1}) + (-283.3\text{ kJ mol}^{-1}) + (-111.3\text{ kJ mol}^{-1})$$
$$-(-283.3\text{ kJ mol}^{-1}) + (-131.2\text{ kJ mol}^{-1}) - (77.1\text{ kJ mol}^{-1}) - (-111.3\text{ kJ mol}^{-1})$$
$$\underline{= -55.7\text{ kJ mol}^{-1}}$$

P9.3) Calculate $\Delta S^{\circ}_{reaction}$ for the reaction $AgNO_3(aq) + KCl(aq) \rightarrow AgCl(s) + KNO_3(aq)$.

In analogy to P9.1, $\Delta S^{\circ}_{reaction}$ is given by:

$$\Delta S^{\circ}_{reaction} = (96.3\text{ J mol}^{-1}) + (102.5\text{ J mol}^{-1}) + (146.4\text{ J mol}^{-1})$$
$$-(102.5\text{ J mol}^{-1}) - (56.5\text{ J mol}^{-1}) - (72.7\text{ J mol}^{-1}) - (146.4\text{ J mol}^{-1}) \underline{= -32.9\text{ J K}^{-1}\text{ mol}^{-1}}$$

P9.7) Express $\mu_{\pm}$ in terms of μ_{+} and μ_{-} for (a) NaCl, (b) $MgBr_2$, (c) Li_3PO_4, and (d) $Ca(NO_3)_2$. Assume complete dissociation.

$\mu_{\pm}$ is given by:

$$\mu_{\pm} = \frac{(\nu_{+}\mu_{+} + \nu_{-}\mu_{-})}{\nu}$$

a) $\mu_{\pm}(\text{NaCl}) = \frac{(\mu_{+} + \mu_{-})}{2}$

b) $\mu_{\pm}(\text{MgBr}_2) = \frac{(\mu_{+} + 2\mu_{-})}{3}$

c) $\mu_{\pm}(\text{Li}_3\text{PO}_4) = \frac{(3\mu_{+} + \mu_{-})}{4}$

d) $\mu_{\pm}(\text{K}_4\text{Fe(CN)}_6) = \frac{(4\mu_{+} + \mu_{-})}{5}$

P9.10) Calculate the ionic strength in a solution that is 0.0050 *m* in K_2SO_4, 0.0010 *m* in Na_3PO_4, and 0.0025 *m* in $MgCl_2$.

The ionic strength, I, is given by:

$$I = \sum_i \frac{m_i}{2}\left(\nu_{i+}\, z_{i+}{}^2 + \nu_{i-}\, z_{i-}{}^2\right)$$

$$I = \frac{(0.005\text{ m})}{2}(2\times1^2 + 1\times2^2) + \frac{(0.001\text{ m})}{2}(3\times1^2 + 1\times3^2) + \frac{(0.0025\text{ m})}{2}(1\times2^2 + 2\times1^2) = \underline{0.0285}$$

P9.14) Calculate the Debye–Hückel screening length $1/\kappa$ at 298 K in a 0.00100 *m* solution of NaCl.

For water, the screening length at 298 K in m^{-1} can be calculated as:

$$\kappa = 9.211\times10^8 \sqrt{\frac{I/(\text{mol kg}^{-1})}{\varepsilon_r}} = 9.211\times10^8 \sqrt{\frac{(0.00100)}{(78.54)}}\ \text{m}^{-1} = 3.29\times10^6\ \text{m}^{-1}$$

$$\frac{1}{\kappa} = 3.04\times10^{-7}\ \text{m} = \underline{304\text{ nm}}$$

P9.17) Calculate *I*, $\gamma_{\pm}$, and $a_{\pm}$ for a 0.0250 *m* solution of $AlCl_3$ at 298 K. Assume complete dissociation.

The ionic strength, I, is given by:

$$I = \frac{m}{2}\left(\nu_+ z_+^{\ 2} + \nu_- z_-^{\ 2}\right) = \frac{(0.025)}{2}\left(1\times 3^2 + 3\times 1^2\right) = \underline{0.15 \text{ mol kg}^{-1}}$$

$\gamma_\pm$ is given by:

$$\gamma_\pm = \text{Exp}\left[-1.173\left|z_+ z_-\right|\sqrt{I}\right] = \text{Exp}\left[-1.173\left|3\times(-1)\right|\sqrt{0.15 \text{ mol kg}^{-1}}\right] = \underline{0.2559}$$

And finally, $a_\pm$ can be obtained:

$$m_\pm = \sqrt[\nu]{m_+^{\ \nu_+}\ m_+^{\ \nu_+}} = \sqrt[4]{(0.025)^1 \times (3\times 0.025)^3} = 0.0570$$

$$a_\pm = m_\pm \gamma_\pm = (0.0570)\times(0.2559) = \underline{0.0146}$$

P9.20) Calculate the solubility of $BaSO_4$ ($K_{sp} = 1.08 \times 10^{-10}$) (a) in pure H_2O and (b) in an aqueous solution with $I = 0.0010$ mol kg^{-1}.

a) With:

$$\nu_+ = 1,\ \nu_- = 1,\ z_+ = 2,\ z_+ = 2,\ c_{SO_4^{\ 2-}} = c_{Ba^{2+}}$$

$$K_{sp} = \left(\frac{c_{Ba^{2+}}}{c^\circ}\right)\left(\frac{c_{SO_4^{\ 2-}}}{c^\circ}\right)\gamma_\pm^2 = 1.08\times 10^{-10}$$

$$K_{sp} = \left(\frac{c_{Ba^{2+}}}{c^\circ}\right)^2 \gamma_\pm^2 = 1.08\times 10^{-10}$$

when $\gamma_\pm = 1$, $c_{Ba^{2+}} = 1.039\times 10^{-5}$ mol L^{-1}

$$I = \frac{m}{2}\left(\nu_+ z_+^{\ 2} + \nu_- z_-^{\ 2}\right) = \frac{\left(1.039\times 10^{-5} \text{ mol L}^{-1}\right)}{2}\left(1\times 2^2 + 1\times 2^2\right) = 4.157\times 10^{-5} \text{ mol kg}^{-1}$$

$$\ln(\gamma_\pm) = -1.173\left|z_+ z_-\right|\sqrt{I} = -1.173\times(4)\times\sqrt{4.157\times 10^{-5} \text{ mol kg}^{-1}} = -0.03025$$

$$\gamma_\pm = 0.97020$$

when $\gamma_\pm = 0.97020$, $c_{Ba^{2+}} = 1.0711\times 10^{-5}$ mol L^{-1}

$$I = \frac{m}{2}\left(\nu_+ z_+^{\ 2} + \nu_- z_-^{\ 2}\right) = \frac{\left(1.039\times 10^{-5} \text{ mol L}^{-1}\right)}{2}\left(1\times 2^2 + 1\times 2^2\right) = 4.2846\times 10^{-5} \text{ mol kg}^{-1}$$

$$\ln(\gamma_\pm) = -1.173\left|z_+ z_-\right|\sqrt{I} = -1.173\times(4)\times\sqrt{4.2846\times 10^{-5} \text{ mol kg}^{-1}} = -0.03071$$

$$\gamma_\pm = 0.9698$$

when $\gamma_\pm = 0.9698$, $c_{Ba^{2+}} = 1.0716\times 10^{-5}$ mol L^{-1}

$$I = \frac{m}{2}\left(\nu_+ z_+^{\,2} + \nu_- z_-^{\,2}\right) = \frac{\left(1.0716\times10^{-5}\ \mathrm{mol\,L^{-1}}\right)}{2}\left(1\times2^2 + 1\times2^2\right) = 4.2866\times10^{-5}\ \mathrm{mol\,kg^{-1}}$$

$$\ln(\gamma_\pm) = -1.173\left|z_+ z_-\right|\sqrt{I} = -1.173\times(4)\times\sqrt{4.2866\times10^{-5}\ \mathrm{mol\,kg^{-1}}} = -0.03072$$

$$\gamma_\pm = 0.9697$$

The solubility is $1.0716\times10^{-5}\ \mathrm{mol\,L^{-1}}$.

b) $I = 0.0010\ \mathrm{mol\,kg^{-1}}$

$$\ln(\gamma_\pm) = -1.173\left|z_+ z_-\right|\sqrt{I} = -1.173\times(4)\times\sqrt{0.0010\ \mathrm{mol\,kg^{-1}}} = -0.148374$$

$$\gamma_\pm = 0.8621$$

Using:

$$K_{sp} = \left(\frac{c_{Ba^{2+}}}{c^\circ}\right)^2 \gamma_\pm^2 = \left(\frac{c_{Ba^{2+}}}{c^\circ}\right)^2 (0.8621)^2 = 1.08\times10^{-10}$$

And solving for the concentration of Ba^{2+} gives:

$$c_{Ba^{2+}} = 1.21\times10^{-5}\ \mathrm{mol\,kg^{-1}}$$

P9.23) The equilibrium constant for the hydrolysis of dimethylamine,

$$(CH_3)_2NH(aq) + H_2O(aq) \rightarrow CH_3NH^+{}_3(aq) + OH^-(aq)$$

is 5.12×10^{-4}. Calculate the extent of hydrolysis for (a) a 0.125 *m* solution of $(CH_3)_2NH$ in water and (b) a solution that is also 0.045 *m* in $NaNO_3$.

a) Using:

$$K = \frac{m^2\gamma_\pm^2}{c - m} = \frac{m^2\gamma_\pm^2}{(0.125\ \mathrm{m}) - m} = 5.12\times10^{-4}$$

With $\gamma_\pm = 1$ we obtain:

$$m = 7.75\times10^{-3}\ \mathrm{mol\,kg^{-1}}$$

when $\gamma_\pm = 1$:

$$I = (2)\frac{m}{2} = m = 7.75\times10^{-3}\ \mathrm{mol\,kg^{-1}}$$

$$\ln(\gamma_\pm) = -1.173\left|z_+ z_-\right|\sqrt{I} = -1.173\times(1)\times\sqrt{7.75\times10^{-3}\ \mathrm{mol\,kg^{-1}}} = -0.1033$$

$$\gamma_\pm = 0.9019$$

when $\gamma_{\pm} = 0.9019$:

$$m = 8.561\times10^{-3}\ \text{mol kg}^{-1}$$

$$I = m = 8.561\times10^{-3}\ \text{mol kg}^{-1}$$

$$\ln(\gamma_{\pm}) = -0.1085$$

$$\gamma_{\pm} = 0.8971$$

when $\gamma_{\pm} = 0.8971$:

$$m = 8.605\times10^{-3}\ \text{mol kg}^{-1}$$

$$I = m = 8.605\times10^{-3}\ \text{mol kg}^{-1}$$

$$\ln(\gamma_{\pm}) = -0.1088$$

$$\gamma_{\pm} = 0.8969$$

when $\gamma_{\pm} = 0.8969$:

$$m = 8.61\times10^{-3}\ \text{mol kg}^{-1}$$

And the extent of hydrolysis:

$$\frac{(8.61\times10^{-3}\ \text{mol kg}^{-1})}{(0.125\ \text{m})}\times100\% = \underline{6.89\%}$$

b) Calculating the ionic strength:

$$I = \frac{m}{2}\left(\nu_+ z_+^2 + \nu_- z_-^2\right) = \frac{(0.025)}{2}\left(1\times1^2 + 1\times1^2\right) = 0.045\ \text{mol kg}^{-1}$$

Add the ionic strength from part a) gives:

$$I_{total} = 0.045\ \text{mol kg}^{-1} + 8.605\times10^{-3}\ \text{mol kg}^{-1} = 0.0536\ \text{mol kg}^{-1}$$

$$\ln(\gamma_{\pm}) = -1.173|z_+ z_-|\sqrt{I} = -1.173\times(1)\times\sqrt{0.0536\ \text{mol kg}^{-1}} = -0.2716$$

$$\gamma_{\pm} = 0.7622$$

$$K = \frac{m^2\gamma_{\pm}^2}{m} = \frac{m^2(0.7622)^2}{(0.125\ \text{m})} = 5.12\times10^{-4}$$

Solving for m yields:

$$m = \underline{0.01006\ \text{mol kg}^{-1}}$$

P9.25) Calculate the mean ionic molality, $m_{\pm}$, in 0.0500 *m* solutions of (a) $Ca(NO_3)_2$,

(b) NaOH, (c) $MgSO_4$, and (d) $AlCl_3$.

Using: $m_{\pm} = \sqrt[v]{m_+^{v_+}\, m_+^{v_+}}$

a) $m_{\pm} = \sqrt[3]{(0.0050)^1 \times (2\times 0.0050)^2} = \underline{0.00794}$

b) $m_{\pm} = \sqrt[2]{(0.0050)^1 \times (0.0050)^1} = \underline{0.005}$

c) $m_{\pm} = \sqrt[2]{(0.0050)^1 \times (0.0050)^1} = \underline{0.005}$

d) $m_{\pm} = \sqrt[4]{(0.0050)^1 \times (3\times 0.0050)^3} = \underline{0.0114}$

P9.27) At 25 °C, the equilibrium constant for the dissociation of acetic acid, K_a, is 1.75 × 10^{-5}. Using the Debye–Hückel limiting law, calculate the degree of dissociation in 0.100 and 1.00 *m* solutions. Compare these values with what you would obtain if the ionic interactions had been ignored.

For 0.100 m:

$$K = \frac{m^2\,\gamma_{\pm}^2}{(c-m)} = \frac{m^2\,\gamma_{\pm}^2}{(0.100\,m)-m} = 1.75\times 10^{-5}$$

With $\gamma_{\pm} = 1$ we obtain:

$$m = 1.314\times 10^{-3}\ \text{mol kg}^{-1}$$

$$I = (2)\frac{m}{2} = m = 1.314\times 10^{-3}\ \text{mol kg}^{-1}$$

$$\ln(\gamma_{\pm}) = -1.173\,|z_+\,z_-|\,\sqrt{I} = -1.173\times(1)\times\sqrt{1.314\times 10^{-3}\ \text{mol kg}^{-1}} = -0.04252$$

$$\gamma_{\pm} = 0.9584$$

when $\gamma_{\pm} = 0.9584$:

$$m = 1.38056\times 10^{-3}\ \text{mol kg}^{-1}$$

$$I = m = 1.38056\times 10^{-3}\ \text{mol kg}^{-1}$$

$$\ln(\gamma_{\pm}) = -0.04343$$

$$\gamma_{\pm} = 0.9575$$

when $\gamma_{\pm} = 0.9575$:

$$m = 1.372\times 10^{-3}\ \text{mol kg}^{-1}$$

$I = m = 1.372 \times 10^{-3}\ \text{mol kg}^{-1}$

$\ln(\gamma_{\pm}) = -0.04345$

$\gamma_{\pm} = 0.9575$

This result converged sufficiently to calculate the extent of hydrolysis:

$$\frac{(1.372 \times 10^{-3}\ \text{mol kg}^{-1})}{(0.100\ \text{m})} \times 100\% = \underline{1.37\%}$$

For 1.00 m:

$$K = \frac{m^2\,\gamma_{\pm}^2}{(c - m)} = \frac{m^2\,\gamma_{\pm}^2}{(1.00\ \text{m}) - m} = 1.75 \times 10^{-5}$$

With $\gamma_{\pm} = 1$ we obtain:

$m = 4.1833 \times 10^{-3}\ \text{mol kg}^{-1}$

$$I = (2)\frac{m}{2} = m = 4.1833 \times 10^{-3}\ \text{mol kg}^{-1}$$

$$\ln(\gamma_{\pm}) = -1.173\,|z_+\,z_-|\,\sqrt{I} = -1.173 \times (1) \times \sqrt{4.1833 \times 10^{-3}\ \text{mol kg}^{-1}} = -0.07587$$

$\gamma_{\pm} = 0.9269$

when $\gamma_{\pm} = 0.9269$:

$m = 4.5130 \times 10^{-3}\ \text{mol kg}^{-1}$

$I = m = 4.5130 \times 10^{-3}\ \text{mol kg}^{-1}$

$\ln(\gamma_{\pm}) = -0.07880$

$\gamma_{\pm} = 0.9242$

when $\gamma_{\pm} = 0.9242$:

$m = 4.5263 \times 10^{-3}\ \text{mol kg}^{-1}$

$I = m = 4.5263 \times 10^{-3}\ \text{mol kg}^{-1}$

$\ln(\gamma_{\pm}) = -0.07892$

$\gamma_{\pm} = 0.9241$

when $\gamma_{\pm} = 0.9241$:

$m = 4.5268 \times 10^{-3}\ \text{mol kg}^{-1}$

$I = m = 4.5268 \times 10^{-3}\ \text{mol kg}^{-1}$

$$\ln(\gamma_{\pm}) = -0.07892$$

$$\gamma_{\pm} = 0.9241$$

This result has converged, and the extent of hydrolysis can be calculated:

$$\frac{(4.5268\times10^{-3}\text{ mol kg}^{-1})}{(1.00\text{ m})}\times 100\% = \underline{0.453\%}$$

If the ionic strengths are ignored we obtain for 0.100 m:

$$K = \frac{m^2}{(c-m)} = \frac{m^2}{(0.100\text{ m})-m} = 1.75\times10^{-5}$$

$$m = \sqrt{1.75\times10^{-6}} = 1.323\times10^{-3}\text{ mol kg}^{-1}$$

$$\frac{(1.323\times10^{-3}\text{ mol kg}^{-1})}{(0.100\text{ m})}\times 100\% = \underline{1.32\%}$$

And for 1.00 m:

$$K = \frac{m^2}{(c-m)} = \frac{m^2}{(1.00\text{ m})-m} = 1.75\times10^{-5}$$

$$m = \sqrt{1.75\times10^{-5}} = 1.314\times10^{-3}\text{ mol kg}^{-1}$$

$$\frac{(1.314\times10^{-3}\text{ mol kg}^{-1})}{(0.100\text{ m})}\times 100\% = \underline{0.418\%}$$

P9.31) Calculate ΔG_r° and the equilibrium constant at 298.15 K for the reaction

$$Hg_2Cl_2(s) \rightarrow 2Hg(l) + Cl_2(g).$$

$$\underline{\Delta G^{\circ}_{\text{reaction}} = -210.7\text{ kJ mol}^{-1}}$$

The equilibrium constant is then:

$$K_{eq} = \text{Exp}\left[\frac{\Delta G^{\circ}_{\text{reaction}}}{R\,T}\right] = \text{Exp}\left[\frac{(-210.7\text{ kJ mol}^{-1})}{(8.314472\text{ J mol}^{-1}\text{ K}^{-1})\times(298.15\text{ K})}\right] = \underline{1.222\times10^{-37}}$$

P9.33) Using half-cell potentials, calculate the equilibrium constant at 298.15 K for the reaction $2H_2O(l) \rightarrow 2H_2(g) + O_2(g)$. Compare your answer with that calculated using Table 9.1. What is the value of E° for the overall reaction that makes the two methods agree exactly?

Finding the half cell reaction:

Reduction: $4\,H^+ + 4\,e^- \longrightarrow 2\,H_2$ $\qquad E_0 = 0\,V$

Oxidation: $2\,O^{2-} \longrightarrow O_2 + 4\,e^-$ $\qquad E_0 = \text{-}1.229\,V$

Then:

$E_{cell} = E_{red} + E_{ox} = \underline{\text{-}1.229\,V}$

And the equilibrium constant:

$$\ln\left(K_{eq}\right) = \frac{n\,F}{R\,T} E_{cell}$$

$$K_{eq} = \mathrm{Exp}\left[\frac{n\,F}{R\,T} E_{cell}\right] = \mathrm{Exp}\left[\frac{4\times\left(96485\,C\,mol^{-1}\right)}{\left(8.314472\,J\,mol^{-1}\,K^{-1}\right)\times\left(298.15\,K\right)} \times \left(\text{-}1.229\,V\right)\right] = \underline{7.99\times10^{-84}}$$

P9.35) For the half-cell reaction $Hg_2Cl_2(s) + 2e^- \rightarrow 2Hg(l) + 2Cl^-(aq)$, $E^\circ = +0.27$ V. Using this result and the data tables in Appendix B, determine $\Delta G_f^\circ\left(Cl^-, aq\right)$

$$\Delta G_R^\circ = -n\,F E^\circ = 2\,\Delta G_f^\circ\left(Cl^-, aq\right) - \Delta G_f^\circ\left(Hg_2Cl_2, s\right)$$

$$\Delta G_f^\circ\left(Cl^-, aq\right) = \frac{-\Delta G_f^\circ\left(Hg_2Cl_2, s\right) - n\,F E^\circ}{2}$$

$$\Delta G_f^\circ\left(Cl^-, aq\right) = \frac{\left(-210.7\,kJ\,mol^{-1}\right) - \left(2\,mol\right)\times\left(96485C\,mol^{-1}\right)\times\left(0.26808\,V\right)}{2} = \underline{-131.2\,kJ\,mol^{-1}}$$

P9.47) The standard half-cell potential for the reaction $O_2(g) + 4H^+(aq) + 4e^- \rightarrow 2H_2O$ is +1.03 V at 298.15 K. Calculate E for a 0.50 molal solution of HCl for $a_{O_2} = 1$ assuming (a)that the a_{H^+} is equal to the molality and (b) using the measured mean ionic activity coefficient for this concentration of $\gamma_\pm = 0.757$. How large is the relative error if the concentrations, rather than the activities, are used?

a) Using for a half cell reaction with $a_\pm(H^+) = m$:

$$E = E_0 - \frac{R\,T}{n\,F}\ln\left(\frac{a_{Red}}{a_{Ox}}\right) = \left(1.03\,V\right) - \frac{\left(8.314472\,J\,mol^{-1}\,K^{-1}\right)\times\left(298.15\,K\right)}{4\times\left(96485\,C\,mol^{-1}\right)} \times \ln\left(\frac{1^4}{0.5^4}\right) = \underline{1.0122\,V}$$

b) In analogy to a) with $a_\pm(H^+) = m_\pm \times \gamma_\pm$:

$$E = E_0 - \frac{R\,T}{n\,F}\ln\left(\frac{a_{Red}}{a_{Ox}}\right) = (1.03\text{ V}) - \frac{(8.314472\text{ J mol}^{-1}\text{ K}^{-1})\times(298.15\text{ K})}{4\times(96485\text{ C mol}^{-1})}\times\ln\left(\frac{1^4}{(0.5\times0.757)^4}\right) = \underline{1.0050\text{ V}}$$

$$\text{error} = \frac{(1.0122\text{ V})}{(1.0050\text{ V})} = \underline{0.0072 = 0.72\%}$$

P9.54) Consider the reaction of pyruvate with NADH form to lactate:

$$CH_3(CO)COO^-(aq) + NADH(aq) + H^+(aq)$$
$$\rightleftarrows CH_3(CHOH)COO^-(aq) + NAD^+(aq)$$

Suppose at equilibrium [pyruvate] = 0.0010 *M*, [lactate] = 0.1000 *M*, [NADH] = 0.0010 *M*, and [NAD$^+$] = 0.25210 *M*. Calculate *K*, *K*′, ΔG°, and $\Delta G^{\circ\prime}$ at pH = 7.

a) The equilibrium constant, K, is given by:

$$K = \frac{\left(\frac{c_{lactate}}{c^\circ_{lactate}}\right)^{v_{lactate}}\left(\frac{c_{NAD^+}}{c^\circ_{NAD^+}}\right)^{v_{NAD^+}}}{\left(\frac{c_{pyruvate}}{c^\circ_{pyruvate}}\right)^{v_{pyruvate}}\left(\frac{c_{NADH}}{c^\circ_{NADH}}\right)^{v_{NADH}}\left(\frac{c_{H^+}}{c^\circ_{H^+}}\right)^{v_{H^+}}}$$

$$= \frac{c_{lactate}\,c_{NAD^+}}{c_{pyruvate}\,c_{NADH}\,10^{-pH}} = \frac{(0.1000\text{ mol L}^{-1})\times(0.2521\text{ mol L}^{-1})}{(0.0010\text{ mol L}^{-1})\times(0.0010\text{ mol L}^{-1})\times(10^{-7}\text{ mol L}^{-1})} = \underline{2.52\times10^{11}}$$

b) With x = -1 for H$^+$ as a reactant at pH = 7, K′ is then given by:

$$K' = K\times10^{7x} = 2.52\times10^{11}\times10^{-7} = \underline{2.52\times10^4}$$

c) ΔG° can be calculated as:

$$\Delta G^\circ = R\,T\ln(K) = (8.314472\text{ J mol}^{-1}\text{ K}^{-1})\times(298.15\text{ K})\times\ln(2.52\times10^{11}) = \underline{-65.1\text{ kJ mol}^{-1}}$$

d) And finally, ΔG°′:

$$\Delta G^{\circ\prime} = R\,T\ln(K') = (8.314472\text{ J mol}^{-1}\text{ K}^{-1})\times(298.15\text{ K})\times\ln(2.52\times10^4) = \underline{-25.1\text{ kJ mol}^{-1}}$$

P9.57) For the oxidation of NADH by hydrogen

$NADH(aq) + H^+(aq) \rightarrow NAD^+(aq) + H_2(g)$

assume $C_{NADH} = 3.2\times10^{-2}\,M$, $c_{NAD^+} = 1.5\times10^{-3}\,M$, and $p_{H_2} = 0.015$ bar and pH = 4.5. Using the data in Table 9.7, calculate the emf and the Gibbs energy change.

The first step is to calculate Q:

$$Q=\frac{\left(\frac{p_{H_2}}{p^{\circ}_{H_2}}\right)^{\nu_{lactate}}\left(\frac{c_{NAD^+}}{c^{\circ}_{NAD^+}}\right)^{\nu_{NAD^+}}}{\left(\frac{c_{NADH}}{c^{\circ}_{NADH}}\right)^{\nu_{NADH}}\left(\frac{c_{H^+}}{c^{\circ}_{H^+}}\right)^{\nu_{H^+}}}c^{\circ}_{H^+}=\frac{p_{H_2}\,c_{NAD^+}}{c_{NADH}\,10^{-pH}}c^{\circ}_{H^+}$$

$$=\frac{\left(0.015\text{ mol L}^{-1}\right)\times\left(0.0015\text{ mol L}^{-1}\right)}{\left(0.032\text{ mol L}^{-1}\right)\times\left(3.1623\times10^{-5}\text{ mol L}^{-1}\right)}\times10^{-7}\text{ mol L}^{-1}=\underline{2.2\times10^{-6}}$$

Next, $E^{0}_{reaction}$ can be obtained from the two half reactions:

Reduction: $2\,H^+ + 2\,e^- \longrightarrow H_2$ $\quad E_0 = -0.421\text{ V}$

Oxidation: $NADH \longrightarrow NAD^+ + H^+ + 2\,e^-$ $\quad E_0 = 0.320\text{ V}$

$E^{0}_{reaction} = -0.101\text{ V}$

$$E_{reaction}=E^{0}_{reaction}-\frac{0.05916}{n}\log(Q)=(-0.101\text{ V})-\frac{0.05916}{2}\log\left(2.2\times10^{-6}\right)=\underline{0.066\text{ V}}$$

And then finally:

$$\Delta G_{reaction}=-n\,F\,E_{reaction}=-2\times\left(96485\text{ C mol}^{-1}\right)\times\left(0.066\text{ V}\right)=\underline{-12.8\text{ kJ mol}^{-1}}$$

P9.58) Referring to Figure 9.16, suppose a volume is divided into two compartments by a semipermeable membrane. In the left-hand "protein compartment" is a solution containing the sodium salt of a protein *P*, NaP, with initial concentration 1.00×10^{-2} *M*. In the right-hand "salt compartment" is a NaCl solution initially at 3.00×10^{-1} *M*. Calculate the equilibrium concentrations of all ionic species in both compartments. Calculate also the net osmotic pressure across the membrane. Assume T = 298 K.

With:

$a=c^{R}_{Cl^-}(\text{initial})=c^{R}_{Na^+}(\text{initial})=0.3\text{ M}$ and $b=c^{L}_{P^-}(\text{initial})=c^{L}_{Na^+}(\text{initial})=0.01\text{ M}$,

the concentrations of all species in the two compartments are:

$$c^{L}_{Na^+}(\text{initial})=\frac{(a+b)^2}{(b+2a)}=\frac{(0.01\text{ M}+0.3\text{ M})^2}{(0.01\text{ M}+2\times0.3)}=\underline{0.1575\text{ M}}$$

$$c^{R}_{Cl^-}(\text{eq})=c^{R}_{Na^+}(\text{eq})=\frac{\left(a^2+ab\right)}{(b+2a)}=\frac{\left(0.3^2\text{ M}^2+0.01\text{ M}\times0.3\text{ M}\right)}{(0.01\text{ M}+2\times0.3)}=\underline{0.1524\text{ M}}$$

$$c_{Cl^-}^{L}(\mathrm{eq}) = \frac{a^2}{(b+2a)} = \frac{0.3^2\,\mathrm{M}^2}{(0.01\,\mathrm{M} + 2\times 0.3\,\mathrm{M})} \underline{= 0.1475\,\mathrm{M}}$$

The net osmotic pressure, π, across the membrane is:

$$\pi = \left[\frac{(b^2 + 2\,a\,b + b^2)}{(b + 2\,a)}\right] R\,T$$

$$= \frac{(0.01^2\,\mathrm{M}^2 + 2\times 0.3\,\mathrm{M}\times 0.01\,\mathrm{M} + 0.01^2\,\mathrm{M}^2)}{(0.01\,\mathrm{M} + 2\times 0.3\,\mathrm{M})}\times(8.314472\,\mathrm{J\,mol^{-1}\,K^{-1}})\times(298\,\mathrm{K}) \underline{= 25.2\,\mathrm{Pa}}$$

P9.60) Suppose a volume is divided into two compartments by a semipermeable membrane, as shown in Figure 9.16. In the left-hand "protein compartment" is a solution containing the sodium salt of a protein *P*, Na_2P, with initial concentration 1.00×10^{-2} *M*. In the right-hand "salt compartment" is a NaCl solution initially at 1.00×10^{-1} *M*. Calculate the equilibrium concentrations of all ionic species in both compartments. Calculate also the Donnan ratio and the junction potential resulting from the Donnan effect. Assume $T = 298$ K.

With:

$$a = c_{Cl^-}^{R}(\mathrm{initial}) = c_{Na^+}^{R}(\mathrm{initial}) = 0.1\,\mathrm{M} \text{ and } b = 2\times c_{Na^+}^{L}(\mathrm{initial}) = 0.02\,\mathrm{M},$$

the concentrations of all species in the two compartments are:

$$c_{Na^+}^{L}(\mathrm{initial}) = \frac{(a+b)^2}{(b+2a)} = \frac{(0.1\,\mathrm{M} + 0.02\,\mathrm{M})^2}{(0.02\,\mathrm{M} + 2\times 0.1)} \underline{= 0.0655\,\mathrm{M}}$$

$$c_{Cl^-}^{R}(\mathrm{eq}) = c_{Na^+}^{R}(\mathrm{eq}) = \frac{(a^2 + ab)}{(b+2a)} = \frac{(0.1^2\,\mathrm{M}^2 + 0.1\,\mathrm{M}\times 0.02\,\mathrm{M})}{(0.02\,\mathrm{M} + 2\times 0.1)} \underline{= 0.0545\,\mathrm{M}}$$

$$c_{Cl^-}^{L}(\mathrm{eq}) = \frac{a^2}{(b+2a)} = \frac{0.1^2\,\mathrm{M}^2}{(0.02\,\mathrm{M} + 2\times 0.1\,\mathrm{M})} \underline{= 0.0454\,\mathrm{M}}$$

The Donnan ratio is:

$$r_D = \frac{(a+b)}{a} = \frac{(0.02\,\mathrm{M} + 0.1\,\mathrm{M})}{0.1\,\mathrm{M}} \underline{= 1.2}$$

The junction potential is finally:

$$\varphi_D = -\frac{R\,T}{F}\ln(r_D) = \frac{(8.314472\,\mathrm{J\,mol^{-1}\,K^{-1}})\times(298.15\,\mathrm{K})}{(96485\,\mathrm{C\,mol^{-1}})}\times\ln(1.2) \underline{= 4.68\,\mathrm{mV}}$$

Chapter 10: Principles of Biochemical Thermodynamics

P10.3) The cellular concentrations of glucose, glucose-6-phosphate, ADP, and ATP are $5.00 \times 10^{-3}\,M$, $8.3 \times 10^{-5}\,M$, $1.38 \times 10^{-4}\,M$, and $1.85 \times 10^{-3}\,M$, respectively. Using the data in Table 10.1, calculate the Gibbs energy change for the phosphorylation of glucose by ATP.

We use:

$$\Delta G = \Delta G^{\circ|}_{reaction} + R\,T\ln Q$$

The reaction of glucose with ATP can be written as the sum of a hydrolysis and phosphorylation:

$$glu + ATP \longrightarrow ADP + glu-6-P \qquad \Delta G^{\circ\prime} = 13.8\,kJ\,mol^{-1}$$

$$ATP + H_2O \longrightarrow ADP + P_i \qquad \Delta G^{\circ\prime} = -30.5\,kJ\,mol^{-1}$$

Therefore:

$$\Delta G^{\circ|}_{reaction} = -30.5\,kJ\,mol + 13.8\,kJ\,mol = -16.7\,kJ\,mol$$

$$R\,T\ln Q = \left(8.314472\,J\,mol^{-1}\,K^{-1}\right)\times\left(310\,K\right)\times\ln\left(\frac{\left(1.38\times10^{-4}\,M\right)\times\left(8.3\times10^{-5}\,M\right)}{\left(1.85\times10^{-3}\,M\right)\times\left(0.005\,M\right)}\right)$$

$$= -17.25\,kJ\,mol^{-1}$$

$$\Delta G = \Delta G^{\circ|}_{reaction} + R\,T\ln Q = -16.7\,kJ\,mol - 17.25\,kJ\,mol = \underline{-33.95\,kJ\,mol}$$

P10.6) Calculate the reversible work required to transport 1mol of K^+ from a region where $C_{K^+} = 5.25$ mM to a region where $C_{K^+} = 35.5$ mM if the potential change accompanying this movement is $\Delta\phi = 0.055$ V. Assume $T = 298$ K.

The reversible work is given by:

$$\Delta\tilde{\mu} = R\,T\ln\left(\frac{c_{out}}{c_{in}}\right) + n\,F\,\Delta\phi$$

$$= \left(8.314472\,J\,mol^{-1}\,K^{-1}\right)\times\left(310\,K\right)\times\ln\left(\frac{(35.5\,mM)}{(5.25\,mM)}\right) + 1\times\left(96485\,C\,mol^{-1}\right)\times\left(0.055\,V\right)$$

$$\underline{= 10.0\,kJ\,mol^{-1}}$$

P10.10) The net reaction for active transport of sodium and potassium ions is thought to be:

$$\left.\begin{matrix} 3Na^+(\text{inside}) \\ + \\ 2K^+(\text{outside}) \end{matrix}\right\} + ATP \rightarrow ADP + \text{phosphate} + \left\{\begin{matrix} 3Na^+(\text{outside}) \\ + \\ 2K^+(\text{outside}) \end{matrix}\right.$$

The concentrations of sodium and potassium ions inside and outside a cell, and the electrical potential E inside and outside the cell are as follows:

	C_{Na^+} (mol L^{-1})	C_{K^+} (mol L^{-1})	ϕ (V)
Outside	1.40×10^{-1}	5.00×10^{-3}	0.00
Inside	1.00×10^{-2}	1.00×10^{-1}	-7.00×10^{-2}

a. Calculate the free energy change involved in transporting 1 mol of sodium ion out of the cell. Assume the activity coefficients of sodium ion inside and outside the cell are unity. Assume the temperature is 310 K.

b. Calculate the free energy change involved in transporting 1 mol of potassium ion into the cell. Assume the activity coefficients of potassium ion inside and outside the cell are unity. Assume the temperature is 310 K.

c. Calculate the total free energy change involved in transporting 3 mol of sodium ion out of the cell and 2mol of potassium into the cell at $T = 310$ K. Assume, as in parts (a) and (b), that all activity coefficients are unity.

a) The free energy change involved in transporting 1 mol of sodium ions out of the cell is given by:

$$\Delta G = RT\ln\left(\frac{c_{\text{out}}^{Na^+}}{c_{\text{in}}^{Na^+}}\right) + F(\phi_{out} - \phi_{in})$$

$$= (8.314472\text{ J mol}^{-1}\text{ K}^{-1}) \times (310\text{ K}) \times \ln\left(\frac{(0.14\text{ M})}{(0.01\text{ M})}\right) + (96485\text{ C mol}^{-1}) \times (0\text{ V} + 7.0\times10^{-2}V) = \underline{13.6\text{ kJ mol}^{-1}}$$

b) The free energy change involved in transporting 1 mol of potassium ions into the cell is given by:

$$\Delta G = RT\ln\left(\frac{c_{\text{in}}^{K^+}}{c_{\text{out}}^{K^+}}\right) + F(\phi_{in} - \phi_{out})$$

$$= (8.314472\text{ J mol}^{-1}\text{ K}^{-1}) \times (310\text{ K}) \times \ln\left(\frac{(0.1\text{ M})}{(0.005\text{ M})}\right) + (96485\text{ C mol}^{-1}) \times (-7.0\times10^{-2}V + 0\text{ V}) = \underline{0.97\text{ kJ mol}^{-1}}$$

c) The free energy change involved in transporting 3 moles of sodium out of the cell and 2 moles of potassium ions into the cell is given by:

$$\Delta G = \mathrm{R\,T}\left(3\ln\left(\frac{c_{\text{out}}^{Na^+}}{c_{\text{in}}^{K^+}}\right)+2\ln\left(\frac{c_{\text{in}}^{K^+}}{c_{\text{out}}^{K^+}}\right)\right)+\mathrm{F}\left(\phi_{out}-\phi_{in}\right)$$

$$=\left(8.314472\ \mathrm{J\ mol^{-1}\ K^{-1}}\right)\times\left(310\ \mathrm{K}\right)\times$$

$$\left(3\ln\left(\frac{(0.14\ \mathrm{M})}{(0.01\ \mathrm{M})}\right)+2\ln\left(\frac{(0.1\ \mathrm{M})}{(0.005\ \mathrm{M})}\right)\right)+\left(96485\ \mathrm{C\ mol^{-1}}\right)\times\left(7.0\times10^{-2}\ \mathrm{V}\right)=\underline{42.60\ \mathrm{kJ\ mol^{-1}}}$$

P10.13) For the hydrolysis $ATP^{4-} + H_2O \rightarrow ADP^{3-} + HPO_4^{2-} + H^+$, $\Delta G^{\circ\prime} = -30.5\ \mathrm{kJ\ mol^{-1}}$ at pH = 7 and T = 310 K. Calculate K', K, and ΔG° for the hydrolysis of ATP^{4-}.

K' for the hydrolysis is given by:

$$\Delta G^{\circ\prime} = -\mathrm{R\,T}\ln K'$$

$$K' = \mathrm{Exp}\left[-\frac{\Delta G^{\circ\prime}}{\mathrm{R\,T}}\right] = \mathrm{Exp}\left[-\frac{\left(-30500\ \mathrm{J\ mol^{-1}}\right)}{\left(8.314472\ \mathrm{J\ mol^{-1}\ K^{-1}}\right)\times\left(310\ \mathrm{K}\right)}\right] = \underline{1.38\times10^{5}}$$

With x = 1 since H^+ ions are on the product side of the reaction:

$$K' = K\,10^{7x}$$

$$K = K'10^{-7} = \left(10^{-7}\right)\times\left(1.38\times10^{5}\right) = \underline{1.38\times10^{-2}}$$

And:

$$\Delta G = -\mathrm{R\,T}\ln K = -\left(8.314472\ \mathrm{J\ mol^{-1}\ K^{-1}}\right)\times\left(310\ \mathrm{K}\right)\times\ln\left(1.38\times10^{-2}\right) = \underline{11.0\ \mathrm{kJ\ mol^{-1}}}$$

P10.15) Calculate K_{obs}, ΔG°_{obs}, and ΔS°_{obs} for ATP hydrolysis at pH 7 and at T = 298.15 K.

The hydrolysis of ATP is:

$$ATP + H_2O \longrightarrow ADP + P_i$$

K_{obs} can be calculated as:

$$K_{obs} = \frac{K_1}{c_{H^+}}\frac{x_{ATP}}{x_{ADP}\,x_P} = \frac{10^{-pK(K_1)}}{c_{H^+}}\frac{x_{ATP}}{x_{ADP}\,x_P} = \frac{(0.628)}{\left(10^{-7}\right)}\times\frac{(0.529)}{(0.568)\times(0.624)} = \underline{9.37\times10^{6}}$$

ΔG°_{obs} is then:

$$\Delta G^\circ_{obs} = -R\,T\ln K_{obs} = -\left(8.314472\text{ J mol}^{-1}\text{ K}^{-1}\right)\times\left(298.15\text{ K}\right)\times\ln\left(9.37\times10^{6}\right)$$

$$\underline{= -39.8\text{ kJ mol}^{-1}}$$

To obtain ΔS°_{obs} we first calculate ΔH°_{obs} :

$$\Delta H^\circ_{obs} = \Delta H^\circ_1 - x_{ADP}\frac{c_{H^+}}{K_{1ADP}}\Delta H^\circ_{1ADP} - x_P\frac{c_{H^+}}{K_{2P}}\Delta H^\circ_{2P} + x_{ATP}\frac{c_{H^+}}{K_{1ATP}}\Delta H^\circ_{1ATP}$$

$$= \left(-19.7\text{ kJ mol}^{-1}\right) - (0.568)\times\frac{\left(10^{-7}\right)}{\left(10^{-6.88}\right)}\times\left(-5.7\text{ kJ mol}^{-1}\right) - (0.624)\times\frac{\left(10^{-7}\right)}{\left(10^{-6.78}\right)}\times\left(3.3\text{ kJ mol}^{-1}\right)$$

$$+ (0.529)\times\frac{\left(10^{-7}\right)}{\left(10^{-6.95}\right)}\times\left(-7.0\text{ kJ mol}^{-1}\right) = -21.78\text{ kJ mol}^{-1}$$

Then:

$$\Delta S^\circ_{obs} = \frac{\left(\Delta G^\circ_{obs} - \Delta H^\circ_{obs}\right)}{T} = \frac{\left(-39.8\text{ kJ mol}^{-1} + 21.78\text{ kJ mol}^{-1}\right)}{\left(298.15\text{ K}\right)} \underline{= -60.4\text{ J mol}^{-1}\text{ K}^{-1}}$$

Chapter 11: Biochemical Equilibria

P11.1) The standard Gibbs energy change for step 1 in Equation (11.16) is $\Delta G^{\circ\prime} = -38.6 \text{ kJ mol}^{-1}$. Using this fact and the data in Table 11.4 calculate the half-cell potential for $FMN + 2e^- + 2H^+ \rightarrow FMNH_2$.

The half-cell potential is calculated from:

$$E^{\circ}_{cell} = E^{\circ}_{red} + E^{\circ}_{ox}$$

The overall reaction is:

$$FMN + FMN + H^+ \longrightarrow NAD^+ + FMNH_2 \text{ with } \Delta G^0 = \text{-38.6 kJ mol}^{-1}$$

giving:

$$E^{\circ}_{cell} = -\frac{\Delta G^0}{n\,F} = -\frac{(-38600 \text{ J mol}^{-1})}{2 \times (96485 \text{ C mol}^{-1})} = 0.20 \text{ V}$$

The oxidation part of the overall reaction is:

$$NADH \longrightarrow NAD + 2\,e^- + H^+ \qquad E^{\circ}_{ox} = 0.32 \text{ V}$$

Then:

$$E^{\circ}_{red} = E^{\circ}_{cell} - E^{\circ}_{ox} + = 0.20 \text{ V} - 0.32 \text{ V} \underline{= -0.12 \text{ V}}$$

P11.5) Consider the reduction of pyruvate to lactate by NADH:

$$CH_3(CO)COO^- + NADH + H^+ \rightleftarrows CH_3\,(CHOH)\,COO^- + NAD^+$$

Suppose at equilibrium the concentrations of reactants and products are $C_{lactate} = 0.00200$ *M*, $C_{pyruvate} = 0.05$ *M*, $C_{NADH} = 0.00200$ *M*, $c_{NAD^-} = 0.504$ *M* . Calculate K, K', $\Delta G^{\circ}, \Delta G^{\circ\prime}$ at pH = 7. Assume $T = 298$ K.

Starting by calculating the equilibrium constant:

$$K = \frac{(0.504) \times (0.002)}{(10^{-7}) \times (0.002) \times (0.05)} \underline{= 1.01 \times 10^8}$$

With x = -1 since H^+ ions are on the reactant side of the reaction:

$$K' = K\,10^{7x}$$

$$K' = K\,10^{-7} = (10^{-7}) \times (1.01 \times 10^8) \underline{= 10.08}$$

Then:

$$\Delta G^{0}_{reaction} = -R\,T\ln K = \left(8.314472\text{ J mol}^{-1}\text{ K}^{-1}\right)\times\left(298\text{ K}\right)\times\ln\left(1.01\times10^{8}\right) = \underline{-45.7\text{ kJ mol}}$$

$$\Delta G^{\circ\prime}_{reaction} = -R\,T\ln K' = \left(8.314472\text{ J mol}^{-1}\text{ K}^{-1}\right)\times\left(298\text{ K}\right)\times\ln\left(10.08\right) = \underline{-5.72\text{ kJ mol}}$$

P11.11) Studies of the binding of ATP to the enzyme tetrahydrofolate synthetase were conducted at T =293 K and appear in the following table, where C_{ATP} is in molar units.

$\bar{\nu}$	0.25	0.50	1.0	1.5	2.0	2.5	3
C_{ATP} *(M)*	6.67×10^{-6}	1.43×10^{-5}	3.33×10^{-5}	6.00×10^{-5}	1.00×10^{-4}	1.67×10^{-4}	3×10^{-4}

a. From a Scatchard plot, determine K_{293} and N.

b. Assume $K_{293}/K_{310} = 2$. Using the information from the Scatchard plot of part (a), calculate the standard enthalpy $\Delta H°$ for the binding of ATP to tetrahydrofolate synthetase. Assume $\Delta H°$ is constant between $T = 293$ K and $T = 310$ K.

c. Calculate $\Delta G°$, the standard Gibbs energy change for the binding of ATP to tetrahydrofolate synthetase, at $T = 293$ K.

d. Calculate $\Delta S°$, the standard entropy change for the binding of ATP to tetrahydrofolate synthetase.

Using the data in the table, plotting $\dfrac{\bar{\nu}}{c_L}$ versus $\bar{\nu}$ yields:

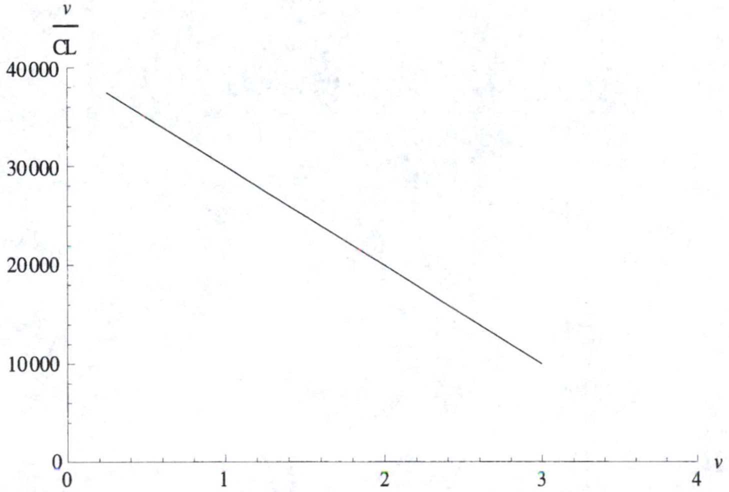

From the equation:

$$\frac{\bar{\nu}}{c_L} = k\left(N - \bar{\nu}\right) = kN - k\bar{\nu}$$

-k is the slope. Therefore:

$$k = -\text{slope} = -\frac{\Delta\frac{\bar{v}}{c_L}}{\Delta\bar{v}} = -\frac{(37481.3-10000)}{(0.25-3.0)} = 9993.2 \ \underline{= 0.9993\times10^4}$$

The $y-\text{axis intercept} = k\,N$, therefore:

$$N = \frac{\text{intercept}}{k} = \frac{40000}{9993.2} \underline{\cong 4}$$

From Chapter 6:

$$\ln K(T_2) = \ln K(T_1) - \frac{\Delta H^\circ_{reaction}}{R}\left(\frac{1}{T_2} - \frac{1}{T_1}\right)$$

Solving for $\Delta H^\circ_{reaction}$ yields:

$$\Delta H^\circ_{reaction} = \frac{R\times(\ln K(T_2) - \ln K(T_1))}{\left(\frac{1}{T_2} - \frac{1}{T_1}\right)} = \frac{(8.314472\ \text{J mol}^{-1}\ \text{K}^{-1})\times\left(\ln K\left(\frac{9993.2}{2}\right) - \ln K(9993.2)\right)}{\left(\frac{1}{(310\,\text{K})} - \frac{1}{(293\,\text{K})}\right)}$$

$$= \underline{-30.82\ \text{kJ mol}^{-1}}$$

$\Delta G^\circ_{reaction}$ is given by:

$$\Delta G^\circ_{reaction} = -R\,T\ln K = -(8.314472\ \text{J mol}^{-1}\ \text{K}^{-1})\times(293\,\text{K})\times\ln(9993.2) \ \underline{= -22.5\ \text{kJ mol}^{-1}}$$

And finally:

$$\Delta S^\circ_{reaction} = \frac{(\Delta G^\circ_{reaction} - \Delta H^\circ_{reaction})}{T} = \frac{(-22.5\ \text{kJ mol}^{-1} + 30.82\ \text{kJ mol}^{-1})}{(293\,\text{K})} \ \underline{= 28.4\ \text{J mol}^{-1}\ \text{K}^{-1}}$$

Chapter 12: From Classical to Quantum Mechanics

P12.2) For a monatomic gas, one measure of the "average speed" of the atoms is the root mean square speed, $v_{rms} = \left|v^2\right|^{1/2} = \sqrt{3kT/m}$, in which m is the molecular mass and k is the Boltzmann constant. Using this formula, calculate the de Broglie wavelength for He and Ar atoms at 100. and at 500. K.

The de Broglie wave lengths can be calculated as follows:

$$\lambda = \frac{h}{p} = \frac{h}{m\,v} == \frac{h}{m\sqrt{\frac{3\,k\,T}{m}}}$$

$$\lambda(\mathrm{He}, 100\,\mathrm{K}) = \frac{\left(6.626\times10^{-34}\ \mathrm{J\,s}\right)}{\left(\frac{\left(4.003\times10^{-3}\ \mathrm{kg\ mol^{-1}}\right)}{\left(6.022\times10^{-23}\ \mathrm{mol^{-1}}\right)}\right)\sqrt{\frac{3\left(1.38066\times10^{-23}\ \mathrm{J\,K^{-1}}\right)\times\left(100\ \mathrm{K}\right)}{\left(\frac{\left(4.003\times10^{-3}\ \mathrm{kg\ mol^{-1}}\right)}{\left(6.022\times10^{-23}\ \mathrm{mol^{-1}}\right)}\right)}}}$$

$$\underline{= 1.263\times10^{-10}\ \mathrm{m}}$$

$$\lambda(\mathrm{He}, 500\,\mathrm{K}) = \frac{\left(6.626\times10^{-34}\ \mathrm{J\,s}\right)}{\left(\frac{\left(4.003\times10^{-3}\ \mathrm{kg\ mol^{-1}}\right)}{\left(6.022\times10^{-23}\ \mathrm{mol^{-1}}\right)}\right)\sqrt{\frac{3\left(1.38066\times10^{-23}\ \mathrm{J\,K^{-1}}\right)\times\left(500\ \mathrm{K}\right)}{\left(\frac{\left(4.003\times10^{-3}\ \mathrm{kg\ mol^{-1}}\right)}{\left(6.022\times10^{-23}\ \mathrm{mol^{-1}}\right)}\right)}}}$$

$$\underline{= 5.65\times10^{-11}\ \mathrm{m}}$$

$$\lambda(\mathrm{Ar}, 100\,\mathrm{K}) = \frac{\left(6.626\times10^{-34}\ \mathrm{J\,s}\right)}{\left(\frac{\left(39.95\times10^{-3}\ \mathrm{kg\ mol^{-1}}\right)}{\left(6.022\times10^{-23}\ \mathrm{mol^{-1}}\right)}\right)\sqrt{\frac{3\left(1.38066\times10^{-23}\ \mathrm{J\,K^{-1}}\right)\times\left(100\ \mathrm{K}\right)}{\left(\frac{\left(39.95\times10^{-3}\ \mathrm{kg\ mol^{-1}}\right)}{\left(6.022\times10^{-23}\ \mathrm{mol^{-1}}\right)}\right)}}}$$

$$\underline{= 4.00\times10^{-11}\ \mathrm{m}}$$

$$\lambda(\mathrm{Ar}, 500\,\mathrm{K}) = \frac{\left(6.626\times10^{-34}\ \mathrm{J\,s}\right)}{\left(\frac{\left(39.95\times10^{-3}\ \mathrm{kg\ mol^{-1}}\right)}{\left(6.022\times10^{-23}\ \mathrm{mol^{-1}}\right)}\right)\sqrt{\frac{3\left(1.38066\times10^{-23}\ \mathrm{J\,K^{-1}}\right)\times\left(500\ \mathrm{K}\right)}{\left(\frac{\left(39.95\times10^{-3}\ \mathrm{kg\ mol^{-1}}\right)}{\left(6.022\times10^{-23}\ \mathrm{mol^{-1}}\right)}\right)}}}$$

$$\underline{= 1.79\times10^{-11}\ \mathrm{m}}$$

P12.4) Electrons have been used to determine molecular structure by diffraction. Calculate the speed of an electron for which the wavelength is equal to a typical bond length, namely, 0.150 nm.

$$v=\frac{p}{m}=\frac{h}{m\lambda}=\frac{\left(6.626\times10^{-34}\ \mathrm{J\,s}\right)}{\left(9.109\times10^{-31}\ \mathrm{kg}\right)\times\left(0.15\times10^{-9}\ \mathrm{m}\right)}=\underline{4.86\times10^{6}\ \mathrm{m\,s^{-1}}}$$

P12.6) Pulsed lasers are powerful sources of nearly monochromatic radiation. Lasers that emit photons in a pulse of 10.-ns duration with a total energy in the pulse of 0.10 J at 1000. nm are commercially available.

a. What is the average power (energy per unit time) in units of watts (1 W = 1 J/s) associated with such a pulse?

b. How many 1000.-nm photons are emitted in such a pulse?

a) The energy per second is given by:

$$E=\frac{\left(0.10\ \mathrm{J}\right)}{\left(10\times10^{-9}\ \mathrm{s}\right)}=\underline{10^{7}\ \mathrm{J\,s^{-1}}=10^{7}\ \mathrm{W}}$$

b) To obtain the number of photons per pulse we first calculate the energy per photon wavelength of 1000.0 nm:

$$E_{photon}=h\nu=h\frac{c}{\lambda}=\left(6.626\times10^{-34}\ \mathrm{J\,s}\right)\times\frac{\left(2.998\times10^{8}\ \mathrm{m\,s^{-1}}\right)}{\left(1000.0\times10^{-9}\ \mathrm{m}\right)}=1.986\times10^{-19}\ \mathrm{J}$$

The number of photons per pulse is then the energy per pulse, calculated in a) divided by the energy per photon:

$$N_{photons}=\frac{E}{E_{photon}}=\frac{\left(0.10\ \mathrm{J}\right)}{\left(1.986\times10^{-19}\ \mathrm{J}\right)}=\underline{5.03\times10^{17}}$$

P12.8) A 1000.-W gas discharge lamp emits 3.00 W of ultraviolet radiation in a narrow range centered near 280. nm. How many photons of this wavelength are emitted per second?

To obtain the number of photons emitted per second we need to calculate the energy per photon with a wavelength of 280.0 nm:

$$E_{photon}=h\nu=h\frac{c}{\lambda}=\left(6.626\times10^{-34}\ \mathrm{J\,s}\right)\times\frac{\left(2.998\times10^{8}\ \mathrm{m\,s^{-1}}\right)}{\left(280.0\times10^{-9}\ \mathrm{m}\right)}=7.095\times10^{-19}\ \mathrm{J}$$

The number of photons is then the energy of the radiation per second divided by the energy per photon:

$$N_{photons} = \frac{E}{E_{photon}} = \frac{(3.00\,J\,s^{-1})}{(7.095\times10^{-19}\,J)} = \underline{4.23\times10^{18}\,s^{-1}}$$

P12.10) What speed does a H_2 molecule have if it has the same momentum as a photon of wavelength 280. nm?

To obtain the speed of the H2 molecule we set the momenta of the H_2 molecule and the photon equal to each other:

$$p_{H_2} = p_{photon}$$

$$m_{H_2}\,v_{H_2} = \frac{h}{\lambda}$$

and solve for the speed, v:

$$v_{H_2} = \frac{h}{\lambda\, m_{H_2}} = \frac{(6.626\times10^{-34}\,J\,s)}{(280.0\times10^{-9}\,m)\times(2\times1.677\times10^{-27}\,kg)} = \underline{0.705\,m\,s^{-1}}$$

P12.13) The power per unit area emitted by a blackbody is given by $P = \sigma T^4$ with $\sigma = 5.67\times10^{-8}\,W\,m^{-2}\,K^{-4}$. Calculate the energy radiated per second by a spherical blackbody of radius 0.500 m at 1000. K. What would the radius of a blackbody at 2500. K be if it emitted the same energy as the spherical blackbody of radius 0.500 m at 1000.K?

a) The power emitted per unit area (m^2) by the blackbody per m^{-2} is:

$$P = \sigma T^4 = (5.67\times10^{-8}\,W\,m^{-2}\,K^{-4})\times(1000\,K)^4 = 56700\times10^{-8}\,W\,m^{-2}$$

The surface of the sphere is given by:

$$S_{sphere} = 4\,\pi\,r^2 = 3.14159\,m^2$$

The power emitted by the blackbody is then:

$$P_{black\,body} = (56700\times10^{-8}\,W\,m^{-2})\times(3.14159\,m^2) = \underline{1.78\times10^5\,W}$$

b) To get the radius of a blackbody that emits the same energy as in a) at 2500 K we use:

$$P = \sigma T^4\,S_{sphere} = \sigma T^4\,4\,\pi\,r^2$$

Solving for r:

$$r = \sqrt{\frac{P}{\sigma T^4 S_{sphere}}} = \sqrt{\frac{(1.78\times10^5\ W)}{(5.67\times10^{-8}\ W\ m^{-2}\ K^{-4})\times(2500\ K)^4\times(3.14159\ m^2)}} = \underline{0.08\ m}$$

P12.14) The power per unit area radiated by blackbody per unit area of surface expressed in units of W m^{-2} is given by $P = \sigma T^4$ with $\sigma = 5.67 \times 10^{-8}$ W m^{-2} K^{-4}. The radius of the sun is 7.00×10^5 km and the surface temperature is 6000. K. Calculate the total energy radiated per second by the sun. Assume ideal blackbody behavior.

The total energy radiation emitted by the sun per second is:

$$P = \sigma T^4 S_{sphere} = \sigma T^4\ 4\pi\ r^2 = (5.67\times10^{-8}\ W\ m^{-2}\ K^{-4})\times(6000\ K)^4\times4\pi\times(7\times10^8\ m)^2$$

$$\underline{= 4.52\times10^{26}\ W}$$

P12.19) The work function of platinum is 5.65 eV. What is the minimum frequency of light required to observe the photoelectric effect on Pt? If light with a 150.-nm wavelength is absorbed by the surface, what is the velocity of the emitted electrons?

The minimum energy corresponds to the work function. First, we convert eV to J:

$$E_{min} = (5.65\,eV)\times\frac{(1.602\times10^{-19}\ J)}{(1\,eV)} = 9.0513\times10^{-19}\ J$$

The minimum frequency is then obtained from:

$$E_{min} = h\,\nu_{min}$$

$$\nu_{min} = \frac{E_{min}}{h} = \frac{(9.0513\times10^{-19}\ J)}{(6.626\times10^{-34}\ J\,s)} = \underline{1.37\times10^{15}\ s^{-1}}$$

To obtain the velocity of the emitted electron we need to calculate the energy:

$$E_{emitted} = E_{photon} - E_{min} = \frac{h\,c}{\lambda} - E_{min}$$

$$= \frac{(6.626\times10^{-34}\ J\,s)\times(2.998\times10^8\ m\,s^{-1})}{(150.0\times10^{-9}\ m)} - 9.0513\times10^{-19}\ J = 4.1919\times10^{-19}\ J$$

The velocity is then obtained as:

$$v_{emitted} = \sqrt{\frac{2\,E_{emitted}}{m}} = \sqrt{\frac{2\times(4.1919\times10^{-19}\ J)}{(9.109\times10^{-31}\ kg)}} = \underline{9.59\times10^5\ m\,s^{-1}}$$

Chapter 13: The Schrödinger Equation

P13.2) Consider a two-level system with $\varepsilon_1 = 3.10 \times 10^{-21}$ J and $\varepsilon_2 = 6.10 \times 10^{-21}$ J. If $g_2 = g_1$, what value of T is required to obtain $n_2/n_1 = 0.150$? What value of T is required to obtain $n_2/n_1 = 0.999$?

Using:

$$\frac{n_2}{n_1} = \frac{g_2}{g_1} \mathrm{Exp}\left[-\frac{(\varepsilon_2 - \varepsilon_1)}{k\,T}\right]$$

Solving for T yields with $n_2/n_1 = 0.150$:

$$T = \frac{(\varepsilon_1 - \varepsilon_2)}{k\left(\ln\frac{n_2}{n_1} + \ln\frac{g_1}{g_2}\right)} = \frac{(3.10\times10^{-21}\,\mathrm{J} - 6.10\times10^{-21}\,\mathrm{J})}{(1.38066\times10^{-23}\,\mathrm{J\,K^{-1}})\times(\ln(0.150) + \ln(1))} = \underline{114.5\,\mathrm{K}}$$

For $n_2/n_1 = 0.999$:

$$T = \frac{(\varepsilon_1 - \varepsilon_2)}{k\left(\ln\frac{n_2}{n_1} + \ln\frac{g_1}{g_2}\right)} = \frac{(3.10\times10^{-21}\,\mathrm{J} - 6.10\times10^{-21}\,\mathrm{J})}{(1.38066\times10^{-23}\,\mathrm{J\,K^{-1}})\times(\ln(0.999) + \ln(1))} = \underline{217.5\times10^{5}\,\mathrm{K}}$$

P13.4) A wave traveling in the z direction is described by the wave function $\Psi(z,t) = A_1\, \mathbf{x} \sin(kz - \omega t + \phi_1) + A_2\, \mathbf{y} \sin(kz - \omega t + \phi_2)$, where **x** and **y** are vectors of unit length along the x and y axes, respectively. Because the amplitude is perpendicular to the propagation direction, $\Psi(z,t)$ represents a transverse wave.

a. What requirements must A_1 and A_2 satisfy for a plane polarized wave in the x-z plane?

b. What requirements must A_1 and A_2 satisfy for a plane polarized wave in the y-z plane?

c. What requirements must A_1 and A_2 and ϕ_1 and ϕ_2 satisfy for a plane polarized wave in a plane oriented at 45° to the x-z plane?

d. What requirements must A_1 and A_2 and ϕ_1 and ϕ_2 satisfy for a circularly polarized wave?

a) $A_1 \neq 0,\ A_2 = 0$

b) $A_1 = 0,\ A_2 \neq 0$

c) $A_1 = A_2 \neq 0,\ \phi_1 = \phi_2$

d) $A_1 = A_2 \neq 0$, $\phi_1 = \phi_2 + \pi/2$ or $\phi_1 = \phi_2 - \pi/2$

P13.7) Express the following complex numbers in the form $re^{i\theta}$.

a. $2-4i$ **c.** $\dfrac{3+i}{4i}$

b. 6 **d.** $\dfrac{8+i}{2-4i}$

In general for an imaginary number $z = x + i\,y$:

$$r = \sqrt{x^2 + y^2}$$

$$\varphi = \mathrm{Cos}^{-1}\left[\frac{x}{r}\right] = \mathrm{Cos}^{-1}\left[\frac{x}{\sqrt{x^2+y^2}}\right]$$

a) $z = 2 + 4i$

$$r = \sqrt{2^2 + (-4)^2} = 2\sqrt{5}$$

$$\varphi = \mathrm{Cos}^{-1}\left[\frac{2}{2\sqrt{5}}\right] = 1.107 \text{ rad}$$

$$\underline{z = 2\sqrt{5}\,\mathrm{Exp}[1.107\,i]}$$

b) $z = 6 + 0i$

$$r = \sqrt{6^2 + 0^2} = 6$$

$$\varphi = \mathrm{Cos}^{-1}\left[\frac{6}{6}\right] = 0 \text{ rad}$$

$$\underline{z = 6\,\mathrm{Exp}[0\,i] = 6}$$

c) $z = \dfrac{(3+i)}{4i} = \dfrac{1}{4} - \dfrac{3}{4}i$

$$r = \sqrt{\left(\frac{1}{4}\right)^2 + \left(\frac{3}{4}\right)^2} = \frac{\sqrt{10}}{4}$$

$$\varphi = \text{Cos}^{-1}\left[\frac{\left(\frac{1}{4}\right)}{\left(\frac{\sqrt{10}}{4}\right)}\right] = 1.249 \text{ rad}$$

$$\underline{z = \frac{\sqrt{5}}{4}\text{Exp}[1.249\,i]}$$

d) $z = \dfrac{(8+i)}{(2-4i)} = \dfrac{3}{5} - \dfrac{17}{10}i$

$$r = \sqrt{\left(\frac{3}{5}\right)^2 + \left(\frac{17}{10}\right)^2} = \frac{\sqrt{13}}{2}$$

$$\varphi = \text{Cos}^{-1}\left[\frac{\left(\frac{3}{5}\right)}{\left(\frac{\sqrt{13}}{2}\right)}\right] = 1.232 \text{ rad}$$

$$\underline{z = \frac{\sqrt{13}}{2}\text{Exp}[1.232\,i]}$$

P13.15) Carry out the following coordinate transformations:

a. Express the point $x = 3$, $y = 2$, and $z = 1$ in spherical coordinates.

b. Express the point $r = 5$, $\theta = \dfrac{\pi}{4}$, and $\phi = \dfrac{3\pi}{4}$ in Cartesian coordinates.

a) To transform from Cartesian to Spherical Coordinates we use (note that the arctangent must be defined suitably so as to take account of the correct quadrant of y/x):

$$r = \sqrt{x^2 + y^2 + z^2} = \sqrt{3^2 + 2^2 + 1^2} = \underline{\sqrt{14}}$$

$$0 - \text{Cos}^{-1}\left[\frac{z}{r}\right] = \text{Cos}^{-1}\left[\frac{1}{\sqrt{14}}\right] \underline{= 1.30 \text{ rad}}$$

$$\varphi = \text{Tan}^{-1}\left[\frac{y}{x}\right] = \text{Tan}^{-1}\left[\frac{2}{3}\right] \underline{= 0.59 \text{ rad}}$$

b) To transform from Spherical Coordinates to Cartesian we use:

$$x = r\ \mathrm{Sin}[\theta]\ \mathrm{Cos}[\varphi] = 5\,\mathrm{Sin}\left[\frac{\pi}{4}\right]\mathrm{Cos}\left[\frac{3\,\pi}{4}\right] \underline{= -2.5}$$

$$y = r\ \mathrm{Sin}[\theta]\ \mathrm{Sin}[\varphi] = 5\,\mathrm{Sin}\left[\frac{\pi}{4}\right]\mathrm{Sin}\left[\frac{3\,\pi}{4}\right] \underline{= 2.5}$$

$$z = r\ \mathrm{Cos}[\theta] = 5\ \mathrm{Cos}\left[\frac{\pi}{4}\right] \underline{= \frac{5}{\sqrt{2}}}$$

Chapter 14: Using Quantum Mechanics on Simple Systems: The Free Particle, the Particle in a Box, and the Harmonic Oscillator

P14.2) Show that the allowed values of the total energy for the free particle, $E = \hbar^2 k^2/2m$, are consistent with the classical result $E = (1/2)mv^2$.

Starting with:

$$k = \frac{2\pi}{\lambda} \text{ and } \lambda = \frac{h}{p}$$

We obtain:

$$k = \frac{2\pi}{\lambda} = \frac{2\pi p}{h} = \frac{p}{\hbar} = \frac{m v}{\hbar}$$

Then:

$$E = \frac{\hbar^2 k^2}{2m} = \frac{\hbar^2 m^2 v^2}{2m\hbar^2} = \underline{\frac{1}{2} m v^2}$$

P14.4) Evaluate the normalization integral for the solutions of the Schrödinger equation for the particle in the box $\psi_n(x) = A \sin(n\pi x/a)$ using the trigonometric identity $\sin^2 y = (1 - \cos 2y)/2$.

$$1 = A^2 \int_0^a \mathrm{Sin}^2\left[\frac{n\pi x}{a}\right] dx$$

Set $y = \dfrac{n\pi x}{a}$ and $dx = \dfrac{a}{n\pi} dy$

$$1 = A^2 \frac{a}{n\pi} \int_0^{n\pi} \mathrm{Sin}^2[y]\,dy = A^2 \frac{a}{n\pi} \int_0^{n\pi} \frac{(1 - \mathrm{Cos}[2y])}{2} dy = A^2 \frac{a}{n\pi}\left[\frac{y}{2} - \frac{\mathrm{Sin}[2y]}{4}\right]_0^{n\pi}$$

$$1 = A^2 \frac{a}{n\pi} n\pi - A^2 \frac{a}{n\pi}(\mathrm{Sin}[n\pi] - \mathrm{Sin}[0]) = \frac{A^2 a}{2}$$

Therefore:

$$\underline{A = \sqrt{\frac{2}{a}}}$$

P14.6) Calculate the probability that a particle in a one-dimensional box of length a is found between 0.31 a and 0.35 a when it is described by the following wave functions:

$$\sqrt{\frac{2}{a}}\,\text{Sin}^2\left[\frac{\pi x}{a}\right]$$

$$\sqrt{\frac{2}{a}}\,\text{Sin}^2\left[\frac{3\pi x}{a}\right]$$

In general, the probability for finding a particle in a certain region is given by:

$$P = \int_{x_1}^{x_2} \psi(x)^2\,dx$$

$$\text{a) } P(0.31\,a \le x \le 0.35\,a) = \frac{2}{a}\int_{0.31\,a}^{0.31\,a} \text{Sin}^2\left[\frac{\pi x}{a}\right]dx$$

Using:

$$\int \text{Sin}^2[cx]\,dx = \frac{x}{2} - \frac{1}{4c}\text{Sin}[2\,c\,x], \text{ with } c = \pi/a:$$

$$P(0.31\,a \le x \le 0.35\,a) = \frac{2}{a}\left[\frac{x}{2} - \frac{a}{4\pi}\text{Sin}\left[\frac{2\pi x}{a}\right]\right]_{0.31\,a}^{0.35\,a}$$

$$P(0.31\,a \le x \le 0.35\,a) = \frac{2}{a}\left\{\left[\frac{0.35\,a}{2} - \frac{a}{4\pi}\text{Sin}\left[\frac{2\pi\,0.35\,a}{a}\right]\right] - \left[\frac{0.31\,a}{2} - \frac{a}{4\pi}\text{Sin}\left[\frac{2\pi\,0.31\,a}{a}\right]\right]\right\}$$

$$P(0.31\,a \le x \le 0.35\,a) = \frac{2}{a}\{[0.175\,a - 0.0644\,a] - [0.155\,a - 0.0740\,a]\}$$

$$P(0.31\,a \le x \le 0.35\,a) = \frac{2}{a}\{0.1106\,a - 0.081\,a\} = \underline{0.0592}$$

$$\text{b) } P(0.31\,a \le x \le 0.35\,a) = \frac{2}{a}\int_{0.31\,a}^{0.31\,a} \text{Sin}^2\left[\frac{3\pi x}{a}\right]dx$$

Again, using:

$$\int \text{Sin}^2[cx]\,dx = \frac{x}{2} - \frac{1}{4c}\text{Sin}[2\,c\,x], \text{ with } c = 3\pi/a:$$

$$P(0.31\,a \le x \le 0.35\,a) = \frac{2}{a}\left[\frac{x}{2} - \frac{a}{12\pi}\text{Sin}\left[\frac{6\pi x}{a}\right]\right]_{0.31\,a}^{0.35\,a}$$

$$P(0.31\,a \le x \le 0.35\,a) = \frac{2}{a}\left\{\left[\frac{0.35\,a}{2} - \frac{a}{12\,\pi}\,\mathrm{Sin}\left[\frac{6\,\pi\,0.35\,a}{a}\right]\right] - \left[\frac{0.31\,a}{2} - \frac{a}{12\,\pi}\,\mathrm{Sin}\left[\frac{6\,\pi\,0.31\,a}{a}\right]\right]\right\}$$

$$P(0.31\,a \le x \le 0.35\,a) = \frac{2}{a}\{[0.175\,a - 0.0082\,a] - [0.155\,a + 0.0113\,a]\}$$

$$P(0.31\,a \le x \le 0.35\,a) = \frac{2}{a}\{0.1668\,a - 0.1663\,a\} = \underline{0.001}$$

P14.8) Is the superposition wave function $\psi(x) = \sqrt{2/a}\left[\sin(n\pi x/a) + \sin(m\pi x/a)\right]$ a solution of the Schrödinger equation for the particle in the box?

Testing if the superposition function fulfills the Schrödinger equation:

$$\frac{d^2\psi(x)}{dx^2} = \frac{-2\,m}{\hbar}\,E\,\psi(x)$$

$$-\frac{\hbar}{2\,m}\,\frac{d^2\psi(x)}{dx^2} = E\,\psi(x)$$

$$\frac{d^2\psi(x)}{dx^2} = \sqrt{\frac{2}{a}}\left(-\frac{m^2\,\pi^2}{a^2}\,\mathrm{Sin}\left[\frac{m\,\pi\,x}{a}\right] - \frac{n^2\,\pi^2}{a^2}\,\mathrm{Sin}\left[\frac{n\,\pi\,x}{a}\right]\right)$$

Therefore, for the superposition function:

$$-\frac{\hbar}{2\,m}\,\frac{d^2\psi(x)}{dx^2} \ne E\,\psi(x)$$

P14.11) What is the solution of the time-dependent Schrödinger equation $\psi(x,t)$ for the solution of the time-independent Schrödinger equation $\psi_4 x = \sqrt{2/a}\,\sin(4\pi x/a)$ in the particle-in-the-box model? Write $\omega = E/\hbar$ explicitly in terms of the parameters of the problem.

The time dependent wave function is obtained from the time-independent wave function as:

$$\Psi(x,t) = \psi(x)\,\mathrm{Exp}\left[-i\,t\left(\frac{E}{\hbar}\right)\right] = \psi(x)\,\mathrm{Exp}[-i\,t\,\omega]$$

$$\Psi_4(x,t) = \sqrt{\frac{2}{a}}\,\mathrm{Sin}\left[\frac{4\,\pi\,x}{a}\right]\mathrm{Exp}\left[-i\,t\left(\frac{h^2\,4^2}{\hbar\,8\,m\,a^2}\right)\right] = \sqrt{\frac{2}{a}}\,\mathrm{Sin}\left[\frac{4\pi\,x}{a}\right]\mathrm{Exp}\left[-i\,t\left(\frac{2\pi\,h^2\,4^2}{h\,8\,m\,a^2}\right)\right]$$

$$\Psi_4(x,t)=\sqrt{\frac{2}{a}}\,\mathrm{Sin}\left[\frac{4\pi x}{a}\right]\mathrm{Exp}\left[-i\,t\left(\frac{4\pi h}{m a^2}\right)\right]$$

P14.20) Two wave functions are distinguishable if they lead to a different probability density. Which of the following wave functions are distinguishable from $\sin\theta + i\cos\theta$?

a. $\sin\theta + i\cos\theta\left(+\frac{\sqrt{2}}{2}+i\frac{\sqrt{2}}{2}\right)$

b. $i(\sin\theta + i\cos\theta)$

c. $-(\sin\theta + i\cos\theta)$

d. $(\sin\theta + i\cos\theta)\left(-\frac{\sqrt{2}}{2}+i\frac{\sqrt{2}}{2}\right)$

e. $\sin\theta - i\cos\theta$

f. $e^{i\theta}$

To obtain the probability density for the wave function $\mathrm{Sin}[\theta]+i\mathrm{Cos}[\theta]$ we need to find the product with its complex conjugate (simplifying $\mathrm{Sin}[\theta]=S, \mathrm{Cos}[\theta]=C, \sqrt{2}/2=R$:

$$(\mathrm{Sin}[\theta]+i\mathrm{Cos}[\theta])\times(\mathrm{Sin}[\theta]-i\mathrm{Cos}[\theta])=\mathrm{Sin}^2[\theta]+\mathrm{Cos}^2[\theta]$$

Now we can form the equivalent products for the wave functions in a) to c) and compare them.

a) $$(\mathrm{Sin}[\theta]+i\mathrm{Cos}[\theta])\left(\frac{\sqrt{2}}{2}+i\frac{\sqrt{2}}{2}\right)\times(\mathrm{Sin}[\theta]-i\mathrm{Cos}[\theta])\left(\frac{\sqrt{2}}{2}-i\frac{\sqrt{2}}{2}\right)=$$
$$(SR+iSR+iCR-CR)\times(SR-iSR-iCR-CR)=\underline{\mathrm{Sin}^2[\theta]+\mathrm{Cos}^2[\theta]}$$

b) $$i(\mathrm{Sin}[\theta]+i\mathrm{Cos}[\theta])\times(-i)(\mathrm{Sin}[\theta]-i\mathrm{Cos}[\theta]) = \underline{\mathrm{Sin}^2[\theta]+2i\,\mathrm{Sin}[\theta]\mathrm{Cos}[\theta]-\mathrm{Cos}^2[\theta]}$$

c) $$-(\mathrm{Sin}[\theta]+i\mathrm{Cos}[\theta])\times-(\mathrm{Sin}[\theta]-i\mathrm{Cos}[\theta]) = \underline{\mathrm{Sin}^2[\theta]+\mathrm{Cos}^2[\theta]}$$

d) $$(\mathrm{Sin}[\theta]+i\mathrm{Cos}[\theta])\left(-\frac{\sqrt{2}}{2}+i\frac{\sqrt{2}}{2}\right)\times(\mathrm{Sin}[\theta]-i\mathrm{Cos}[\theta])\left(-\frac{\sqrt{2}}{2}-i\frac{\sqrt{2}}{2}\right)=$$
$$(-SR+iSR-iCR-CR)\times(-SR-iSR+iCR-CR) = \underline{\mathrm{Sin}^2[\theta]+\mathrm{Cos}^2[\theta]}$$

That means the wave function from c) is the only wave function distinguishable from $\mathrm{Sin}[\theta]+i\mathrm{Cos}[\theta]$.

P14.26) The force constant for a $H^{19}F$ molecule is 966 N m^{-1}.

a. Calculate the zero point vibrational energy for this molecule for a harmonic potential.

b. Calculate the light frequency needed to excite this molecule from the ground state to the first excited state.

a) To zero point energy for the $H^{19}F$ molecule can be obtained from with n = 0:

$$E_n = \hbar\sqrt{\frac{k}{\mu}}\left(n+\frac{1}{2}\right) = \hbar\sqrt{\frac{k}{\left(\dfrac{m_H\, m_F}{(m_H + m_F)}\right)}}\left(n+\frac{1}{2}\right)$$

$$E_0 = \frac{1}{2}\left(1.0456\times10^{-34}\text{ J s}\right)\sqrt{\frac{\left(966\text{ N m}^{-1}\right)}{\left(\dfrac{\left(1.6772\times10^{-27}\text{ kg}\right)\times\left(3.1551\times10^{-26}\text{ kg}\right)}{\left(\left(1.6772\times10^{-27}\text{ kg}\right)+\left(3.1551\times10^{-26}\text{ kg}\right)\right)}\right)}} = \underline{4.07175\times10^{-20}\text{ J}}$$

b) To calculate the frequency to excite the transition from n = 0 to n = 1 we need to calculate the energy difference between the two states:

$$E_1 = \frac{3}{2}\left(1.0456\times10^{-34}\text{ J s}\right)\sqrt{\frac{\left(966\text{ N m}^{-1}\right)}{\left(\dfrac{\left(1.6772\times10^{-27}\text{ kg}\right)\times\left(3.1551\times10^{-26}\text{ kg}\right)}{\left(\left(1.6772\times10^{-27}\text{ kg}\right)+\left(3.1551\times10^{-26}\text{ kg}\right)\right)}\right)}} = 1.22152\times10^{-19}\text{ J}$$

$$\nu_{0\to1} = \frac{(E_1 - E_0)}{h} = \frac{\left(\left(1.22152\times10^{-19}\text{ J}\right)-\left(4.07175\times10^{-20}\text{ J}\right)\right)}{\left(6.5697\times10^{-34}\text{ J s}\right)} = \underline{1.24\times10^{14}\text{ s}^{-1}}$$

P14.28) Show by carrying out the appropriate integration that the solutions of the Schrödinger equation for the harmonic oscillator $\psi_0 x = (\alpha/\pi)^{1/4} e^{-(1/2)\alpha x^2}$ and $\psi_2(x) = (\alpha/4\pi)^{1/4}\left(2\alpha x^2 - 1\right)e^{-(1/2)\alpha x^2}$ are orthogonal over the interval $-\infty < x < \infty$ and that $\psi_2(x)$ is normalized over the same interval. In evaluating integrals of this type, $\int_{-x}^{x} f(x)dx = 0$ if $f(x)$ is an odd function of x and $\int_{-x}^{x} f(x)dx = 2\int_0^{x} f(x)dx$ if $f(x)$ is an even function of x.

To show that the wave functions are orthogonal, the following integral has to be zero. So for $\psi_0(x)$ and $\psi_2(x)$:

$$\int_{-\infty}^{\infty}\psi_0(x)^*\psi_2(x)dx=\left(\frac{\alpha}{\pi}\right)^{1/4}\left(\frac{\alpha}{4\pi}\right)^{1/4}\int_{-\infty}^{\infty}e^{-\frac{1}{2}\alpha x^2}\left(2\alpha x^2-1\right)e^{-\frac{1}{2}\alpha x^2}dx=0$$

Since the functions are even, the lower integration limit can be set to zero while the integral value is multiplied by 2:

$$\int_{-\infty}^{\infty}\psi_0(x)^*\psi_2(x)dx=2\left(\frac{\alpha}{\pi}\right)^{1/4}\left(\frac{\alpha}{4\pi}\right)^{1/4}\int_{0}^{\infty}e^{-\alpha x^2}\left(2\alpha x^2-1\right)dx$$

$$\int_{-\infty}^{\infty}\psi_0(x)^*\psi_2(x)dx=2\left(\frac{\alpha}{\pi}\right)^{1/4}\left(\frac{\alpha}{4\pi}\right)^{1/4}\left\{\int_{0}^{\infty}2\alpha x^2e^{-\alpha x^2}dx-\int_{0}^{\infty}e^{-\alpha x^2}dx\right\}$$

$$\int_{-\infty}^{\infty}\psi_0(x)^*\psi_2(x)dx=2\left(\frac{\alpha}{\pi}\right)^{1/4}\left(\frac{\alpha}{4\pi}\right)^{1/4}\left\{\frac{\pi^{1/2}}{2\alpha^{1/2}}-\frac{\pi^{1/2}}{2\alpha^{1/2}}\right\}\underline{=0}$$

And for $\psi_2(x)$ to be normalized we need:

$$\int_{-\infty}^{\infty}\psi_2(x)^*\psi_2(x)dx=\left(\frac{\alpha}{4\pi}\right)^{1/2}\int_{-\infty}^{\infty}\left(2\alpha x^2-1\right)e^{-\frac{1}{2}\alpha x^2}\left(2\alpha x^2-1\right)e^{-\frac{1}{2}\alpha x^2}dx=1$$

$$\int_{-\infty}^{\infty}\psi_2(x)^*\psi_2(x)dx=2\left(\frac{\alpha}{4\pi}\right)^{1/2}\int_{0}^{\infty}e^{-\alpha x^2}\left(2\alpha x^2-1\right)^2dx$$

$$\int_{-\infty}^{\infty}\psi_0(x)^*\psi_2(x)dx=2\left(\frac{\alpha}{4\pi}\right)^{1/2}\left\{\int_{0}^{\infty}4\alpha^2x^4e^{-\alpha x^2}dx-\int_{0}^{\infty}4\alpha x^2e^{-\alpha x^2}dx+\int_{0}^{\infty}e^{-\alpha x^2}dx\right\}$$

$$\int_{-\infty}^{\infty}\psi_0(x)^*\psi_2(x)dx=2\left(\frac{\alpha}{4\pi}\right)^{1/2}\left\{\frac{3\pi^{1/2}}{2\alpha^{3/2}}-\frac{\pi^{1/2}}{\alpha^{1/2}}+\frac{\pi^{1/2}}{2\alpha^{1/2}}\right\}=\frac{2\alpha^{1/2}}{2\pi^{1/2}}\left\{\frac{\pi^{1/2}}{\alpha^{1/2}}\right\}\underline{=1}$$

P14.33) The vibrational frequency of $^1H^{35}Cl$ is $8.963\times10^{13}\ s^{-1}$. Calculate the force constant of the molecule. How large a mass would be required to stretch a classical spring with this force constant by 1.00 cm? Use the gravitational acceleration on Earth at sea level for this problem.

The force constant can be obtained using:

$$\omega=\sqrt{\frac{k}{\mu}}$$

$$k = \mu\,\omega^2 = \left(\frac{m_H\, m_{Cl}}{(m_H + m_{Cl})}\right)(2\pi)^2\,\nu^2$$

$$= \left(\frac{\left(1.6772\times10^{-27}\ \text{kg}\right)\times\left(5.81202\times10^{-26}\ \text{kg}\right)}{\left(\left(1.6772\times10^{-27}\ \text{kg}\right)+\left(5.81202\times10^{-26}\ \text{kg}\right)\right)}\right)\times(2\pi)^2\times\left(8.963\times10^{13}\ \text{s}^{-1}\right)^2 = \underline{517.0\ \text{kg s}^{-2}}$$

To get the mass required to stretch the spring 1.00 cm we use the potential energy of a mass in the earth's gravitational field and set it equal to the energy of a mass on the spring:

$$V(x)^{\text{earth}} = V(x)^{\text{spring}}$$

$$m\,g\,x = \frac{1}{2}k\,x^2$$

Solving for m yields:

$$m = \frac{1}{2}\frac{k\,x}{g} = \frac{1}{2}\frac{\left(517.0\ \text{kg s}^{-2}\right)\times\left(0.01\ \text{m}\right)}{\left(9.80665\ \text{m s}^{-2}\right)} = \underline{0.264\text{kg}}$$

P14.35) Use $\sqrt{\langle x^2\rangle}$ as calculated in Problem P14.32 as a measure of the vibrational amplitude for a molecule. What fraction is $\sqrt{\langle x^2\rangle}$ of the 127-pm bond length of the HCl molecule for n = 0, 1, and 2? The force constant for the $^1H^{35}Cl$ molecule is 516 N m^{-1}.

For n = 0:

$$\langle x^2\rangle = \int_{-\infty}^{\infty}\psi_0(x)^*\,x^2\,\psi_0(x)\,dx = \int_{-\infty}^{\infty}x^2\,\psi_0(x)^2\,dx = 2\int_0^{\infty}\left(\frac{\alpha}{\pi}\right)^{1/2}x^2 e^{-\alpha x^2}\,dx$$

$$\langle x^2\rangle = 2\int_0^{\infty}\left(\frac{\alpha}{\pi}\right)^{1/2}x^2 e^{-\alpha x^2}\,dx = \frac{2\,\pi^{1/2}\,\alpha^{1/2}}{4\,\alpha^{3/2}\,\pi^{1/2}} = \frac{1}{2\alpha}$$

With $\alpha = \frac{\hbar}{\sqrt{k\,\mu}}$ we obtain:

$$\sqrt{\langle x^2\rangle} = \sqrt{\frac{\hbar}{2\sqrt{k\,\mu}}} = \sqrt{\frac{\left(1.0456\times10^{-34}\ \text{J s}\right)}{2\sqrt{\left(516\ \text{N m}^{-1}\right)\times\left(1.63014\times10^{-27}\ \text{kg}\right)}}} = 7.55004\times10^{-12}\ \text{m}$$

And the fraction of the 127-pm H-Cl bond is:

$$\frac{7.55004\times10^{-12}\ \text{m}}{127\times10^{-12}\ \text{m}} \underline{=\ 0.059}$$

For n = 1:

$$\langle x^2\rangle = \int_{-\infty}^{\infty}\psi_1(x)^* x^2\psi_1(x)dx = \int_{-\infty}^{\infty}x^2\psi_1(x)^2dx = 2\left(\frac{4\alpha^3}{\pi}\right)^{1/2}\int_0^{\infty}x^4e^{-\alpha x^2}dx$$

$$\langle x^2\rangle = 2\left(\frac{4\alpha^3}{\pi}\right)^{1/2}\int_0^{\infty}x^4e^{-\alpha x^2}dx = 2\left(\frac{4\alpha^3}{\pi}\right)^{1/2}\frac{3\pi^{1/2}}{8\alpha^{5/2}} = \frac{3}{2\alpha}$$

With $\alpha = \frac{\hbar}{\sqrt{k\mu}}$ we obtain:

$$\sqrt{\langle x^2\rangle} = \sqrt{\frac{3\hbar}{2\sqrt{k\mu}}} = \sqrt{\frac{3\times(1.0456\times10^{-34}\ \text{J s})}{2\sqrt{(516\ \text{N m}^{-1})\times(\ 1.63014\times10^{-27}\ \text{kg})}}} = 1.3077\times10^{-11}\ \text{m}$$

And the fraction of the 127-pm H-Cl bond is:

$$\frac{1.3077\times10^{-11}\ \text{m}}{127\times10^{-12}\ \text{m}} \underline{=\ 0.103}$$

For n = 2:

$$\langle x^2\rangle = \int_{-\infty}^{\infty}\psi_2(x)^* x^2\psi_2(x)dx = \int_{-\infty}^{\infty}x^2\psi_1(x)^2dx = 2\left(\frac{\alpha}{4\pi}\right)^{1/2}\int_0^{\infty}x^2\left(2\alpha x^2-1\right)^2 e^{-\alpha x^2}dx$$

$$\langle x^2\rangle = 2\left(\frac{\alpha}{4\pi}\right)^{1/2}\left\{\int_0^{\infty}4\alpha^2x^4e^{-\alpha x^2}dx - 2\int_0^{\infty}2\alpha x^2e^{-\alpha x^2}dx + \int_0^{\infty}x^2e^{-\alpha x^2}dx\right\}$$

$$\langle x^2\rangle = 2\left(\frac{\alpha}{4\pi}\right)^{1/2}\left\{\frac{15\pi^{1/2}}{4\alpha^{3/2}} - 2\frac{3\pi^{1/2}}{4\alpha^{3/2}} + \frac{\pi^{1/2}}{4\alpha^{3/2}}\right\} = \frac{5}{2\alpha}$$

With $\alpha = \frac{\hbar}{\sqrt{k\mu}}$ we obtain:

$$\sqrt{\langle x^2\rangle} = \sqrt{\frac{5\hbar}{2\sqrt{k\mu}}} = \sqrt{\frac{5\times(1.0456\times10^{-34}\ \text{J s})}{2\sqrt{(516\ \text{N m}^{-1})\times(\ 1.63014\times10^{-27}\ \text{kg})}}} = 1.68824\times10^{-11}\ \text{m}$$

And the fraction of the 127-pm H-Cl bond is:

$$\frac{1.68824\times10^{-11}\ \text{m}}{127\times10^{-12}\ \text{m}} \underline{=\ 0.133}$$

Chapter 15: The Hydrogen Atom and Many-Electron Atoms

P15.1) Assume that the radius of a hydrogen atom is 25pm and that the nuclear radius is 1.2×10^{-15} m. (a) What fraction of the atomic volume is occupied by the nucleus? (b) What fraction of the atomic mass is due to the nucleus?

a) The fraction of the atomic volume can be calculated from:

$$\frac{V_{nucleus}}{V_{atom}} = \frac{\frac{4}{3}\pi r_{nucleus}^3}{\frac{4}{3}\pi r_{atom}^3} = \frac{r_{nucleus}^3}{r_{atom}^3} = \frac{\left(7.23823\times10^{-45}\ \text{m}\right)^3}{\left(6.54498\times10^{-32}\ \text{m}\right)^3} = \underline{1.11\times10^{-13}}$$

b) The fraction of the atomic mass of the nucleus is:

$$\frac{m_{nucleus}}{m_{atom}} = \frac{m_{nucleus}}{(m_{nucleus} + m_{electron})} = \frac{\left(1.6726\times10^{-27}\ \text{kg}\right)}{\left(1.6726\times10^{-27}\ \text{kg} + 9.109\,10^{-32}\ \text{kg}\right)} = \underline{0.9995}$$

P15.4) Show that the total energy eigenfunctions $\psi_{100}(r)$ and $\psi_{210}(r)$ are orthogonal.

To show that the energy eigenfunctions are orthogonal, we have to evaluate the following integral:

$$\int \psi_{100}^* \, \psi_{210} \, d\tau$$

This normalization integral becomes:

$$\int \psi_{100}^* \, \psi_{210} \, d\tau = N_{100}\, N_{210} \int_0^{\pi}\int_0^{2\pi}\int_0^{\infty} e^{-r/a_0} \frac{r}{a_0} e^{-r/a_0} \text{Cos}[\theta] r^2 \,\text{Sin}[\theta] \,dr\, d\theta\, d\varphi$$

$$\int \psi_{100}^* \, \psi_{210} \, d\tau = N_{100}\, N_{210} \left\{ \int_0^{2\pi} d\varphi \int_0^{\pi} \text{Cos}[\theta]\text{Sin}[\theta]\, d\theta \int_0^{\infty} \frac{r^3}{a_0} e^{-r/a_0}\, dr \right\}$$

The entire integral is zero since:

$$\int_0^{\pi} \text{Cos}[\theta]\text{Sin}[\theta]\, d\theta - \left[-\frac{1}{2} Cos[\theta] \right]_0^{\pi} = \left(-\frac{1}{2} + \frac{1}{2} \right) = 0$$

P15.10) Using the results from Problem 15.9, calculate the probability of finding the electron in the 1 s state outside a sphere of radius 0.5 a_0, 3 a_0, and 5 a_0.

The result from P15.9 b) for the probability density for finding an electron within a sphere of radius r is:

$$1-e^{-2r/a_0}-\frac{2r}{a_0}\left(1+\frac{r}{a_0}\right)e^{-2r/a_0}$$

Therefore, the probability density for finding an electron outside a sphere of radius r is:

$$1-\left(1-e^{-2r/a_0}-\frac{2r}{a_0}\left(1+\frac{r}{a_0}\right)e^{-2r/a_0}\right)1-e^{-2r/a_0}-\frac{2r}{a_0}\left(1+\frac{r}{a_0}\right)e^{-2r/a_0}$$

Evaluating this equation for $0.5a_0$, $3a_0$, and $5a_0$ gives 0.920, 0.0620, and 2.77×10^{-3}, respectively.

P15.11) Show that the function $2\left(1/a_0\right)^{3/2}e^{-r/a_0}$ is a solution of the following differential equation for $l = 0$:

$$R(r)-\frac{\hbar^2}{2\mu r^2}\frac{d}{dr}\left[r^2\frac{dR(r)}{dr}\right]+\left[\frac{\hbar^2 l(l+1)}{2\mu r^2}-\frac{e^2}{4\pi\varepsilon_0 r}\right]R(r)$$

$$= ER(r)$$

What is the eigenvalue? Using this result, what is the value for the principal quantum number n for this function?

To show that the function is a solution we need:

$$R(r)=N\,\mathrm{Exp}[-r/a_0]$$

$$\frac{dR(r)}{dr}=-\frac{N}{a_0}\mathrm{Exp}[-r/a_0]$$

$$\frac{d}{dr}\left[r^2\frac{dR(r)}{dr}\right]=-\frac{2rN}{a_0}\mathrm{Exp}[-r/a_0]+\frac{r^2N}{a_0^2}\mathrm{Exp}[-r/a_0]$$

Then with $\boldsymbol{l}=0$:

$$N\,\mathrm{Exp}[-r/a_0]-\frac{\hbar^2}{2\mu r^2}\left(-\frac{2rN}{a_0}\mathrm{Exp}[-r/a_0]+\frac{r^2N}{a_0^2}\mathrm{Exp}[-r/a_0]\right)-\frac{e^2N}{4\pi\varepsilon_0 r}\mathrm{Exp}[-r/a_0]=E\,R(r)$$

$$N\,\mathrm{Exp}[-r/a_0]+\frac{\hbar^2N}{\mu r a_0}\mathrm{Exp}[-r/a_0]-\frac{\hbar^2N}{2\mu a_0^2}\mathrm{Exp}[-r/a_0]-\frac{e^2N}{4\pi\varepsilon_0 r}\mathrm{Exp}[-r/a_0]=E\,R(r)$$

$$\left(1+\frac{\hbar^2}{\mu\, r\, a_0}-\frac{\hbar^2}{2\mu\, a_0^{\;2}}-\frac{e^2}{4\pi\,\varepsilon_0\, r}\right) \mathrm{N\,Exp}[-r/a_0] = \mathrm{E\,N\,Exp}[-r/a_0]$$

That means that the energy, E, is the eigenvalue:

$$E=\left(1+\frac{\hbar^2}{\mu\, r\, a_0}-\frac{\hbar^2}{2\mu\, a_0^{\;2}}-\frac{e^2}{4\pi\,\varepsilon_0\, r}\right)$$

P15.14) The wave function for the ground state of the He^+ atom is given by $\frac{1}{\sqrt{\pi}}\left(\frac{2}{a_0}\right)^{3/2} e^{-2r/a_0}$. Calculate the mean value of the radius $\langle r\rangle$ at which you would find the 1*s* electron in He.

To calculate $\langle r\rangle$ we must evaluate the following integral:

$$\int \psi^*\langle r\rangle \psi\, d\tau$$

The integral becomes:

$$\int \psi^*\langle r\rangle \psi\, d\tau = N^2 \int_0^{\pi}\int_0^{2\pi}\int_0^{\infty} e^{-2r/a_0}\, r\, e^{-2r/a_0} \mathrm{Cos}[\theta]\, r^2\, \mathrm{Sin}[\theta]\, dr\, d\theta\, d\varphi$$

$$\int \psi^*\langle r\rangle \psi\, d\tau = N^2\left\{\int_0^{2\pi} d\varphi \int_0^{\pi} \mathrm{Cos}[\theta]\, d\theta \int_0^{\infty} r^3\, e^{-4r/a_0}\, dr\right\}$$

$$\int \psi^*\langle r\rangle \psi\, d\tau = N^2\left\{4\pi\frac{6\, a_0^{\;4}}{256}\right\} = \frac{8}{\pi\, a_0^{\;3}}\,\frac{4\pi\, 6\, a_0^{\;4}}{256} = \underline{\frac{3}{4} a_0}$$

P15.19) The *d* orbitals have the nomenclature $d_{z2}, d_{xy}, d_{xz}, d_{yz}$, and $d_{x^2-y^2}$. Show how the *d* orbital

$$\psi_{3d_{yz}}(r,\theta,\phi) = \frac{\sqrt{2}}{81\sqrt{\pi}}\left(\frac{1}{a_0}\right)^{3/2} \frac{r^2}{a_0^2} e^{-r/3a_0} \sin\theta \cos\theta \sin\phi$$

can be written in the form $yzF(r)$.

For the d-orbital wave function:

$$\psi_{3d_{yz}} = \frac{\sqrt{2}}{81\sqrt{\pi}}\left(\frac{1}{a_0}\right)^{3/2}\frac{r^2}{a_0^2}e^{-r/3a_0}\,\mathrm{Sin}[\theta]\mathrm{Cos}[\theta]\mathrm{Sin}[\varphi] = N\,r^2 e^{-r/3a_0}\,\mathrm{Sin}[\theta]\mathrm{Cos}[\theta]\mathrm{Sin}[\varphi]$$

with $N = \frac{\sqrt{2}}{81\sqrt{\pi}}\left(\frac{1}{a_0}\right)^{3/2}\frac{1}{a_0^2}$

a coordinate transformation from spherical to Cartesian coordinates:

$y = r\,\mathrm{Sin}[\theta]\mathrm{Sin}[\varphi]$

$z = r\,\mathrm{Cos}[\theta]$

can be used to express the orbital as:

$$\underline{\psi_{3d_{yz}} = N\,e^{-r/3a_0}\,y\,z = N\,F(r)\,y\,z}$$

P15.21) The diameter of the Earth is 12,756.3 km and the radius of an iron atom is 156 pm. You make a one-dimensional chain of iron atoms equal in length to the Earth's diameter. (a) How many atoms do you need to make the chain? (b) What is the mass of the chain in grams?

a) The number of iron atoms can be calculated as:

$$\mathrm{num} = \frac{d_{earth}}{d_{Fe}} = \frac{d_{earth}}{2r_{Fe}} = \frac{(12756.3\times10^3\,\mathrm{m})}{2\times(156\times10^{-12}\,\mathrm{m})} = \underline{4.09\times10^{16}\,\mathrm{atoms}}$$

b) The mass of the chain would be:

$$m_{chain} = \frac{M_{Fe}\,\mathrm{num}}{N_A} = \frac{(55.85\,\mathrm{g\,mol^{-1}})(4.09\times10^{16}\,\mathrm{atoms})}{(6.022\times10^{23}\,\mathrm{atoms\,mol^{-1}})} = \underline{3.79\times10^{-6}\,\mathrm{g}}$$

P15.27) What is the shortest wavelength of light that Johannes Balmer expected to see emitted from the hydrogen discharge lamp based on his formula $\frac{1}{\lambda} = R_H\left(\frac{1}{2^2} - \frac{1}{n^2}\right)$, $n = 3, 4, 5, \ldots$?

The shortest wavelength corresponds to the highest frequency, and equivalently to the highest energy. Therefore, we need to use $n = \infty$:

$$\lambda = \frac{1}{R_H\left(\frac{1}{2^2}-\frac{1}{n^2}\right)} = \frac{1}{R_H\left(\frac{1}{4}-\frac{1}{\infty^2}\right)} = \frac{1}{R_H\left(\frac{1}{4}\right)} = \frac{4}{\left(109677.581\text{cm}^{-1}\right)} = \underline{3.64705\times10^{-7}\ \text{m} = 364.705\ \text{nm}}$$

P15.33) An approximate formula for the energy levels in a multielectron atom is $E_n \approx -Z_{eff}^2 e^2 / 8\pi\varepsilon_0 a_0 n^2$, $n = 1, 2, 3, \ldots$, where Z_{eff} is the effective nuclear charge felt by an electron in a given orbital. Calculate values for Z_{eff} from the first ionization energies for the elements Li through Ne (SEE www.webelements.com). Compare these values for Z_{eff} with those listed in Table 15.1. How well do they compare?

Using:

$$Z_{eff} = \sqrt{\frac{8\pi E_{ion} n^2 \varepsilon_0 a_0}{e^2 N_A}} = \sqrt{\frac{8\pi E_{ion} 1^2(8.854187810^{-12}\ \text{C}^2\ \text{J}^{-1}\ \text{m}^{-1})\times(5.29177210^{-11}\ \text{m})}{(1.60217710^{-19}\ \text{C})^2\times(6.02210^{23}\ \text{mol}^{-1})}},$$

the effective charges based on the **first** ionization energies for the elements Li to Ne are calculated as:

Element	Li	Be	B	C	N	O	F	Ne
$E_{ionization}$ in kJ mol^{-1}	520.2	899.5	800.6	1086.5	1402.3	1313.9	1681.0	2080.7
Z_{eff}	1.26	1.66	1.56	1.82	2.07	2.00	2.26	2.52

A comparison with the data in Table 15.1 shows that the approximation reproduces the effective charges reasonably well for the second-row elements with only 2s electrons, however, fails to predict the charges for elements with 2p electrons.

Chapter 16: Chemical Bonding in Diatomic Molecules

P16.4) Although He_2 is not a stable molecule, He_2^+ is stable. Explain this difference using the molecular configurations and bond order of the two species.

The electron configuration for He_2^+ is:

$$(1\sigma_g)^2(1\sigma_u^*)^1$$

The bond order, bo, is therefore:

$$bo = \frac{1}{2}(\text{number of bonding electrons} - \text{number of anti bonding electrons}) = \frac{1}{2}(2-1) = \underline{\frac{1}{2}}$$

As a comparison, the electron configuration and bond order for He_2 are, respectively:

$$(1\sigma_1)^2(1\sigma_1^*)^2$$

$$bo = \frac{1}{2}(2-2) = \underline{0}$$

explaining why He_2^+ is stable, whereas He_2 is not.

P16.8) Write the molecular configurations of the following homonuclear diatomic molecules.

a. N_2^+ **b.** N_2^- **c.** O_2^+ **d.** O_2^-

a) N_2^+ : $(1\sigma_g)^2(1\sigma_u^*)^2(2\sigma_g)^2(2\sigma_u^*)^2(1\pi_u)^2(1\pi_u)^2(3\sigma_g)^1$

b) N_2^- : $(1\sigma_g)^2(1\sigma_u^*)^2(2\sigma_g)^2(2\sigma_u^*)^2(1\pi_u)^2(1\pi_u)^2(3\sigma_g)^2(1\pi_u^*)^1$

c) O_2^+ : $(1\sigma_g)^2(1\sigma_u^*)^2(2\sigma_g)^2(2\sigma_u^*)^2(3\sigma_g)^2(1\pi_u)^2(1\pi_u)^2(1\pi_u^*)^1$

d) O_2^- : $(1\sigma_g)^2(1\sigma_u^*)^2(2\sigma_g)^2(2\sigma_u^*)^2(3\sigma_g)^2(1\pi_u)^2(1\pi_u)^2(1\pi_u^*)^2(1\pi_u^*)^1$

P16.11) An electron in the highest occupied MO in the following molecules is excited to the lowest unoccupied MO. For each molecule, specify whether the bond length increases or decreases. Explain your reasoning.

a. Li_2 **b.** N_2 **c.** Be_2

We can use the bond order, bo:

$$bo = \frac{1}{2}(\text{number of bonding electrons} - \text{number of anti bonding electrons})$$

to identify if a bond length gets shorter or longer upon excitation. An increase and decrease in bond order correspond to a decrease and increase in bond length, respectively.

a) Li_2 : $(1\sigma_g)^2(1\sigma_u^*)^2(2\sigma_g)^2 \xrightarrow{\text{excitation}} (1\sigma_g)^2(1\sigma_u^*)^2(2\sigma_g)^1(2\sigma_u^*)^1$

That means that the bond order changes as:

$$bo = \frac{1}{2}(6-4) = 1 \xrightarrow{\text{excitation}} bo = \frac{1}{2}(3-3) = 0,$$

therefore the molecules dissociates.

b) N_2 :

$$(1\sigma_g)^2(1\sigma_u^*)^2(2\sigma_g)^2(2\sigma_u^*)^2(1\pi_u)^2(1\pi_u)^2(3\sigma_g)^2$$
$$\xrightarrow{\text{excitation}} (1\sigma_g)^2(1\sigma_u^*)^2(2\sigma_g)^2(2\sigma_u^*)^2(1\pi_u)^2(1\pi_u)^2(3\sigma_g)^1(1\pi_u^*)^1$$

That means that the bond order changes as:

$$bo = \frac{1}{2}(10-4) = 3 \xrightarrow{\text{excitation}} bo = \frac{1}{2}(9-5) = 2,$$

therefore the bond length increases.

c) Be_2 :

$$(1\sigma_g)^2(1\sigma_u^*)^2(2\sigma_g)^2(2\sigma_u^*)^2 \xrightarrow{\text{excitation}} (1\sigma_g)^2(1\sigma_u^*)^2(2\sigma_g)^2(2\sigma_u^*)^1(1\pi_u)^1$$

That means that the bond order changes as:

$$bo = \frac{1}{2}(4-4) = 0 \xrightarrow{\text{excitation}} bo = \frac{1}{2}(5-3) = 1,$$

therefore Be_2 is not stable, whereas Be_2^* is.

P16.13) Consider the molecules F_2 and F_2^+.

a. Write the molecular configuration for each molecule.

b. Specify the bond order for each molecule.

c. Is either of these molecules paramagnetic?

d. Which of the molecules has the greater bond strength?

e. Which of the molecules has the longer bond length?

a) $F_2 : (1\sigma_g)^2 (1\sigma_u^*)^2 (2\sigma_g)^2 (2\sigma_u^*)^2 (3\sigma_g)^2 (1\pi_u)^2 (1\pi_u)^2 (1\pi_u^*)^2 (1\pi_u^*)^2$

$F_2^+ : (1\sigma_g)^2 (1\sigma_u^*)^2 (2\sigma_g)^2 (2\sigma_u^*)^2 (3\sigma_g)^2 (1\pi_u)^2 (1\pi_u)^2 (1\pi_u^*)^2 (1\pi_u^*)^1$

b) The bond order is defined as:

$$bo = \frac{1}{2}(\text{number of bonding electrons} - \text{number of anti bonding electrons})$$

$$F_2 : \ bo = \frac{1}{2}(10-8) = 1$$

$$F_2^+ : \ bo = \frac{1}{2}(10-7) = 1.5$$

c) F_2^+ is paramagnetic since it has one unpaired electron.

d) The molecule with the higher bond order, F_2^+, has the greater bond strength.

e) The molecule with the lower bond order, F_2, has the longer bond length.

P16.15) Write the molecular configurations of the following heteronuclear diatomic molecules.

a. CN **b.** OH **c.** BF **d.** NH

a) $CN : (1\sigma)^2 (2\sigma)^2 (3\sigma)^2 (4\sigma)^2 (1\pi)^2 (5\sigma)^2 (2\pi)^1$

b) $OH : (1\sigma)^2 (2\sigma)^2 (3\sigma)^2 (4\sigma)^2 (1\pi)^1$

c) $BF : (1\sigma)^2 (2\sigma)^2 (3\sigma)^2 (4\sigma)^2 (1\pi)^2 (5\sigma)^2$

d) $NH : (1\sigma)^2 (2\sigma)^2 (3\sigma)^2 (4\sigma)^2$

Chapter 17: Molecular Structure and Energy Levels for Polyatomic Molecules

P17.2) SiF_4 has four ligands and one lone pair on the central S atom. Which of the following two structures do you expect to be the equilibrium form? Explain your reasoning.

The answer is b), since this structure puts the lone pair the furthest away from the ligands and bonds.

P17.4) Use the VSEPR method to predict the structures of the following:

a. SCl_2

b. BCl_3

a) The S atom in SCl_2 is sp^3 hybridized. Therefore, the S atom has two lone pairs and two ligands, and the molecule is tetrahedral.

b) The B atom in BCl_3 is sp^3 hybridized. Therefore, the B atom has an empty hybridized orbital and three ligands, and the molecule is trigonal planar.

P17.7) Show that the set of orthonormal *sp*-hybrid orbitals that are oriented 180° apart is

$$\psi_a = \frac{1}{\sqrt{2}}\left(\phi_{2s} + \phi_{2p_z}\right) \text{ and } \psi_b = \frac{1}{\sqrt{2}}\left(\phi_{2s} - \phi_{2p_z}\right)$$

The appropriate linear combinations of carbon AOs for the two sp hybridized orbitals are:

$\psi_a = c_1\,\phi_{2s} + c_2\,\phi_{2p_z}$

$\psi_b = c_3\,\phi_{2s} + c_4\,\phi_{2p_z}$

The 2s AO orbitals are spherical symmetrical and contribute equally to the hybrid orbitals, therefore:

$c_1 = c_2$

We set $c_1 < 0$ to make the 2s AO orbital have a positive amplitude in the binding region, and with $\psi_b = \sqrt{c_1^{\,2} + c_3^{\,2}} = 1$:

$$c_1 = c_3 = -\sqrt{\frac{1}{2}}$$

With equivalent arguments for the $2p_z$ AOs we obtain:

$$c_2 = -\sqrt{\frac{1}{2}} \text{ and } c_4 = \sqrt{\frac{1}{2}}$$

And finally:

$$\psi_a = \sqrt{\frac{1}{2}}\left(-\phi_{2s} - \phi_{2p_z}\right)$$

$$\psi_b = \sqrt{\frac{1}{2}}\left(-\phi_{2s} + \phi_{2p_z}\right)$$

P17.11) Show that the set of four equivalent tetrahedrally oriented hybrid orbitals for sp^3 hybridization that are oriented 109.4° is

$$\psi_a = \frac{1}{2}\left(-\phi_{2s} + \phi_{2p_x} + \phi_{2p_y} + \phi_{2p_z}\right)$$

$$\psi_b = \frac{1}{2}\left(-\phi_{2s} - \phi_{2p_x} - \phi_{2p_y} + \phi_{2p_z}\right)$$

$$\psi_c = \frac{1}{2}\left(-\phi_{2s} + \phi_{2p_x} - \phi_{2p_y} - \phi_{2p_z}\right)$$

$$\psi_d = \frac{1}{2}\left(-\phi_{2s} - \phi_{2p_x} + \phi_{2p_y} - \phi_{2p_z}\right)$$

We assume the signs for the 2s orbital wave functions negative to make the 2s AO orbital have a positive amplitude in the binding region for the sp^3 hybrid orbitals. For a tetrahedral arrangement the signs of the three atomic 2p wave functions can be obtained by assigning the signs of the loops of the sp^3 hybrid orbitals (red) based on their orientations relative to the signs of the 3 atomic p-orbitals (x,y,z, blue):

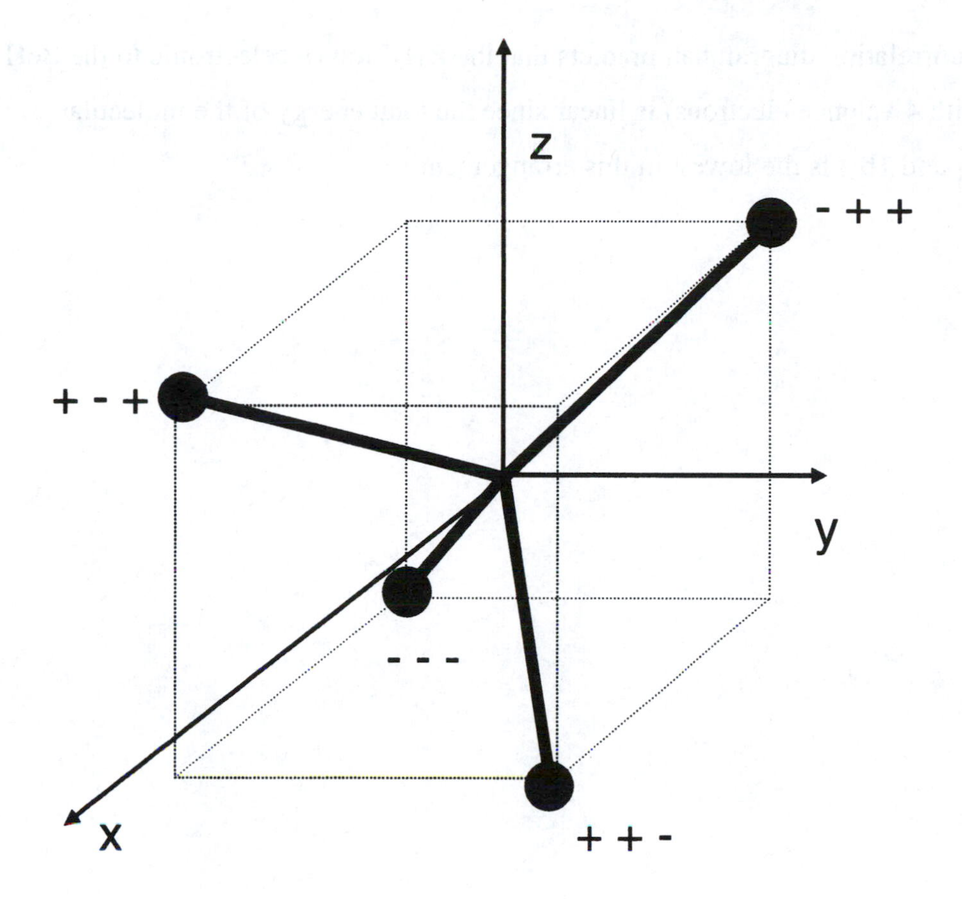

$$\psi_a = \frac{1}{2}\left(-\phi_{2s} + \phi_{2p_x} + \phi_{2p_y} + \phi_{2p_z}\right)$$

$$\psi_b = \frac{1}{2}\left(-\phi_{2s} - \phi_{2p_x} - \phi_{2p_y} + \phi_{2p_z}\right)$$

$$\psi_c = \frac{1}{2}\left(-\phi_{2s} + \phi_{2p_x} - \phi_{2p_y} + \phi_{2p_z}\right)$$

$$\psi_d = \frac{1}{2}\left(-\phi_{2s} - \phi_{2p_x} + \phi_{2p_y} - \phi_{2p_z}\right)$$

P17.14) Use Bent's rule to estimate the change in the H—C—H bond angle when going from H_2CO to H_2CS.

The H—X—H (X = O, S) angle is larger in SCH_2 than in OCH_2, since oxygen is more electronegative than sulfur.

P17.16) Predict whether BH_2^+ is linear or bent using the Walsh correlation diagram in Figure 17.11 Explain your answers.

The Walsh correlation diagramhan predicts that the BH_2^+ ion (isoelectronic to the BeH_2 molecule with 4 valence electrons) is linear since the total energy of the molecular orbitals ($1a_1$ and $1b_2$) is the lowest in this arrangement.

Chapter 18: Vibrational and Rotational Spectroscopy

P18.1) A strong absorption band in the infrared region of the electromagnetic spectrum is observed at $\tilde{\nu} = 2170\ \text{cm}^{-1}$ for $^{12}C^{16}O$. Assuming that the harmonic potential applies, calculate the fundamental frequency ν in units of inverse seconds, the vibrational period in seconds, and the zero point energy for the molecule in joules and electron-volts.

To calculate the fundamental frequency we use:

$$c = \lambda \nu$$

$$\nu = \frac{c}{\lambda} = c\,\tilde{\nu} = \left(2.998\times10^{8}\ \text{m s}^{-1}\right)\times\left(2170\,\text{cm}^{-1}\right)\times\left(\frac{100\,\text{cm}}{1\,\text{m}}\right) = \underline{6.51\times10^{13}\ \text{s}^{-1}}$$

The period is:

$$T = \frac{1}{\nu} = \frac{1}{\left(6.51\times10^{13}\ \text{s}^{-1}\right)} = \underline{1.54\times10^{-14}\ \text{s}^{-1}}$$

And finally, the zero point energy with n = 0 is:

$$E_n = h\,\nu\left(n+\frac{1}{2}\right)$$

$$E_0 = \frac{1}{2}h\,\nu = \frac{1}{2}\left(6.626\times10^{-34}\ \text{J s}^{-1}\right)\times\left(6.51\times10^{13}\ \text{s}^{-1}\right) = \underline{2.16\times10^{-20}\ \text{J}}$$

$$E_0 = \frac{(1\,\text{eV})}{\left(1.602\times10^{-19}\ \text{J}\right)}\left(2.16\times10^{-20}\ \text{J}\right) = \underline{(0.134\,\text{eV})}$$

P18.3) The $^{1}H^{35}Cl$ molecule can be described by a Morse potential with $D_e = 7.41 \times 10^{-19}$ J. The force constant k for this molecule is 516.3 N m^{-1} and $\nu = 8.97 \times 10^{13}\ \text{s}^{-1}$.

a. Calculate the lowest four energy levels for a Morse potential using the following formula:

$$E_n = h\nu\left(n+\frac{1}{2}\right) - \frac{(h\nu)^2}{4D_e}\left(n+\frac{1}{2}\right)^2$$

b. Calculate the fundamental frequency ν_0 corresponding to the transition $n = 0 \rightarrow n = 1$ and the frequencies of the first three overtone vibrations. How large would the relative error be if you assume that the first three overtone frequencies are $2\nu_0$, $3\nu_0$, and $4\nu_0$?

To calculate the population ratios from the Boltzmann equation:

$$\frac{n_a}{n_0} = \mathrm{Exp}\left[-\frac{(E_a - E_0)}{k\,T}\right]$$, with a = 1, 2

we need to determine the energy differences between the states:

$$E_1 - E_0 = h\,\nu\left(1+\frac{1}{2}\right) - h\,\nu\left(0+\frac{1}{2}\right) = h\,\nu = \hbar\sqrt{\frac{k_f}{\mu}}$$

$$E_2 - E_0 = h\,\nu\left(2+\frac{1}{2}\right) - h\,\nu\left(0+\frac{1}{2}\right) = 2\,h\,\nu = 2\,\hbar\sqrt{\frac{k_f}{\mu}}$$

Then:

$$\frac{n_1}{n_0}(H_2, 300\,\mathrm{K}) = \mathrm{Exp}\left[-\frac{\hbar\sqrt{\frac{k_f}{\mu}}}{k\,T}\right] = \mathrm{Exp}\left[-\frac{\hbar\sqrt{\frac{k_f(m_1+m_2)}{m_1\,m_2}}}{k\,T}\right] = \mathrm{Exp}\left[-\frac{\hbar\sqrt{\frac{k_f(m_1+m_2)}{m_1\,m_2}}}{k\,T}\right]$$

$$= \mathrm{Exp}\left[-\frac{(1.0546\times10^{-34}\,\mathrm{J\,s^{-1}})\sqrt{\frac{(575\,\mathrm{N\,m^{-1}})(1.677\times10^{-27}\,\mathrm{kg}+1.677\times10^{-27}\,\mathrm{kg})}{(1.677\times10^{-27}\,\mathrm{kg})\times(1.677\times10^{-27}\,\mathrm{kg})}}}{(1.3807\times10^{-23}\,\mathrm{J\,K^{-1}})\times(300\,\mathrm{K})}\right] = \underline{6.98\times10^{-10}}$$

$$\frac{n_2}{n_0}(H_2, 300\,\mathrm{K}) =$$

$$\mathrm{Exp}\left[-\frac{2\times(1.0546\times10^{-34}\,\mathrm{J\,s^{-1}})\sqrt{\frac{(575\,\mathrm{N\,m^{-1}})(1.677\times10^{-27}\,\mathrm{kg}+1.677\times10^{-27}\,\mathrm{kg})}{(1.677\times10^{-27}\,\mathrm{kg})\times(1.677\times10^{-27}\,\mathrm{kg})}}}{(1.3807\times10^{-23}\,\mathrm{J\,K^{-1}})\times(300\,\mathrm{K})}\right] = \underline{4.87\times10^{-19}}$$

$$\frac{n_1}{n_0}(H_2,1000\,K)=$$

$$\text{Exp}\left[-\frac{\left(1.0546\times10^{-34}\ \text{J s}^{-1}\right)\sqrt{\dfrac{\left(575\ \text{N m}^{-1}\right)\left(1.677\times10^{-27}\ \text{kg}+1.677\times10^{-27}\ \text{kg}\right)}{\left(1.677\times10^{-27}\ \text{kg}\right)\times\left(1.677\times10^{-27}\ \text{kg}\right)}}}{\left(1.3807\times10^{-23}\ \text{J K}^{-1}\right)\times\left(1000\ \text{K}\right)}\right]=\underline{1.79\times10^{-3}}$$

$$\frac{n_2}{n_0}(H_2,1000\,K)=$$

$$\text{Exp}\left[-\frac{2\times\left(1.0546\times10^{-34}\ \text{J s}^{-1}\right)\sqrt{\dfrac{\left(575\ \text{N m}^{-1}\right)\left(1.677\times10^{-27}\ \text{kg}+1.677\times10^{-27}\ \text{kg}\right)}{\left(1.677\times10^{-27}\ \text{kg}\right)\times\left(1.677\times10^{-27}\ \text{kg}\right)}}}{\left(1.3807\times10^{-23}\ \text{J K}^{-1}\right)\times\left(1000\ \text{K}\right)}\right]=\underline{3.21\times10^{-3}}$$

$$\frac{n_1}{n_0}(Br_2,300\,K)=$$

$$\text{Exp}\left[-\frac{\left(1.0546\times10^{-34}\ \text{J s}^{-1}\right)\sqrt{\dfrac{\left(246\ \text{N m}^{-1}\right)\left(1.33\times10^{-25}\ \text{kg}+1.33\times10^{-25}\ \text{kg}\right)}{\left(1.33\times10^{-25}\ \text{kg}\right)\times\left(1.33\times10^{-25}\ \text{kg}\right)}}}{\left(1.3807\times10^{-23}\ \text{J K}^{-1}\right)\times\left(300\ \text{K}\right)}\right]=\underline{0.212}$$

$$\frac{n_2}{n_0}(Br_2,300\,K)=$$

$$\text{Exp}\left[-\frac{2\times\left(1.0546\times10^{-34}\ \text{J s}^{-1}\right)\sqrt{\dfrac{\left(246\ \text{N m}^{-1}\right)\left(1.33\times10^{-25}\ \text{kg}+1.33\times10^{-25}\ \text{kg}\right)}{\left(1.33\times10^{-25}\ \text{kg}\right)\times\left(1.33\times10^{-25}\ \text{kg}\right)}}}{\left(1.3807\times10^{-23}\ \text{J K}^{-1}\right)\times\left(300\ \text{K}\right)}\right]=\underline{0.045}$$

$$\frac{n_1}{n_0}(Br_2, 1000\ K) =$$

$$Exp\left[-\frac{(1.0546\times10^{-34}\ J\ s^{-1})\sqrt{\dfrac{(246\ N\ m^{-1})(1.33\times10^{-25}\ kg + 1.33\times10^{-25}\ kg)}{(1.33\times10^{-25}\ kg)\times(1.33\times10^{-25}\ kg)}}}{(1.3807\times10^{-23}\ J\ K^{-1})\times(1000\ K)}\right] = \underline{0.628}$$

$$\frac{n_2}{n_0}(Br_2, 1000\ K) =$$

$$Exp\left[-\frac{2\times(1.0546\times10^{-34}\ J\ s^{-1})\sqrt{\dfrac{(246\ N\ m^{-1})(1.33\times10^{-25}\ kg + 1.33\times10^{-25}\ kg)}{(1.33\times10^{-25}\ kg)\times(1.33\times10^{-25}\ kg)}}}{(1.3807\times10^{-23}\ J\ K^{-1})\times(1000\ K)}\right] = \underline{0.394}$$

P18.6) Show that the Morse potential approaches the harmonic potential for small values of the vibrational amplitude. (*Hint:* Expand the Morse potential in a Taylor–Mclaurin series.)

The Morse Potential is given by:

$$V(x) = D_e(1 - Exp[-\alpha(x - x_e)])^2$$

and the derivatives:

$$V'(x) = -2\,\alpha\,D_e\ Exp[-\alpha\,(x - x_e)](Exp[-\alpha\,(x - x_e)] - 1)^2$$

$$V''(x) = 2\,\alpha^2\,D_e Exp[-\alpha\,(x - x_e)](2\,Exp[-\alpha\,(x - x_e)] - 1)^2$$

A series expansion into a McLaurin around $(x - x_e)$ yields:

$$f(x) = f(0) + xf'(0) + \frac{x^2}{2!}f''(0) + \frac{x^3}{3!}f'''(0) + \ldots$$

$$V(x) = D_e\,\alpha^2(x - x_e)^2 - D_e\,\alpha^3(x - x_e)^3 + \frac{7}{12}D_e\,\alpha^4(x - x_e)^4 - \ldots$$

For small vibrational amplitudes and with $k = V''(x) = 2\,\alpha^2\,D_e$ the Morse Potential is equivalent to the harmonic potential:

$$V(x) = \frac{1}{2}k\,(x - x_e)^2$$

P18.8) The fundamental vibrational frequencies for $^1H^{19}F$ and $^2D^{19}F$ are 4138.52 and 2998.25 cm^{-1}, respectively, and D_e for both molecules is 5.86 eV. What is the difference in the bond energy of the two molecules?

The bond energy is defined as the energy difference between the first excited state and the ground state:

$$E_{bond} = E_1 - E_0 = h\nu\left(1+\frac{1}{2}\right) - \frac{(h\nu)^2}{4D_e}\left(1+\frac{1}{2}\right)^2 - h\nu\left(0+\frac{1}{2}\right) - \frac{(h\nu)^2}{4D_e}\left(0+\frac{1}{2}\right)^2$$

$$E_{bond}\left({}^1H^{19}F\right) =$$

$$\left(1.0546\times10^{-34}\ J\,s^{-1}\right)\times\left(413852\ m^{-1}\right)\times(1.5) - \frac{\left(1.0546\times10^{-34}\ J\,s^{-1}\right)^2\times\left(413852\ m^{-1}\right)^2}{4\times(5.86\ eV)\times\left(1.6022\times10^{-19}\ J\,eV^{-1}\right)}\times(1.5)$$

$$-\left(1.0546\times10^{-34}\ J\,s^{-1}\right)\times\left(413852\ m^{-1}\right)\times(0.5) - \frac{\left(1.0546\times10^{-34}\ J\,s^{-1}\right)^2\times\left(413852\ m^{-1}\right)^2}{4\times(5.86\ eV)\times\left(1.6022\times10^{-19}\ J\,eV^{-1}\right)}\times(0.5)$$

$$= \underline{2.74218\times10^{-28}\ J}$$

$$E_{bond}\left(2H^{19}F\right) = \left(1.0546\times10^{-34}\ J\,s^{-1}\right)\times\left(299825\ m^{-1}\right)\times(1.5) - \frac{\left(1.0546\times10^{-34}\ J\,s^{-1}\right)^2\times\left(299825\ m^{-1}\right)^2}{4\times(5.86\ eV)\times\left(1.6022\times10^{-19}\ J\,eV^{-1}\right)}\times(1.5)$$

$$-\left(1.0546\times10^{-34}\ J\,s^{-1}\right)\times\left(413852\ m^{-1}\right)\times(0.5) - \frac{\left(1.0546\times10^{-34}\ J\,s^{-1}\right)^2\times\left(413852\ m^{-1}\right)^2}{4\times(5.86\ eV)\times\left(1.6022\times10^{-19}\ J\,eV^{-1}\right)}\times(0.5)$$

$$= \underline{1.60887\times10^{-28}\ J}$$

The difference in bond energy is 1.13331×10^{-28} J.

P18.9) Derive a general relationship between the absorbance and the transmittance using the Beer–Lambert law.

The relationship between the absorbance, A, and transmittance, T, is:

$$A(\lambda) = \ln\left[\frac{I(\lambda)}{I_0(\lambda)}\right] = \ln[T]$$

P18.11) For a 10.0-mm-thick piece of fused silica quartz glass, 50.% of the light incident on it passes through the glass. What percentage of the light will pass through a 20.0-mm-thick piece of the same glass?

Using Beer-Lambert's law:

$$\ln\left[\frac{I(\lambda)}{I_0(\lambda)}\right] = -\varepsilon(\lambda)M\ell$$

setting:

$\ln R_1 = -\varepsilon(\lambda)M\ell_1$ and $\ln R_2 = -\varepsilon(\lambda)M\ell_2$ with:

$R_1 = 0.5$, $\ell_1 = 10$ mm, $R_2 = ?$, $\ell_2 = 20$ mm

$\varepsilon(\lambda)M$ is constant and can be obtained from $\ln R_1 = -\varepsilon(\lambda)M\ell_1$ as:

$$\varepsilon(\lambda)M = -\frac{\ln[R_1]}{\ell_1}$$

therefore, one can obtain R_2 by evaluating:

$$\frac{R_2}{R_1} = Exp\left[\frac{\ln[R_1]}{\ell_1}\ell_2\right]Exp\left[-\frac{\ln[R_1]}{\ell_1}\ell_1\right] = Exp\left[\ln[R_1]\frac{\ell_2}{\ell_1} - \ln[R_1]\right]$$

$$R_2 = R_1 Exp\left[\ln[R_1]\frac{\ell_2}{\ell_1} - \ln[R_1]\right]$$

$$R_2 = 0.5Exp\left[\ln[0.5]\frac{20mm}{10mm} - \ln[0.5]\right]$$

$$\underline{R_2 = 0.25}$$

P18.15) Purification of water for drinking using UV light is a viable way to provide potable water in many areas of the world. Experimentally, the decrease in UV light of wavelength 250 nm follows the empirical relation $I/I_0 = e^{-\varepsilon' l}$ where l is the distance that the light passed through the water and ε' is an effective absorption coefficient. $\varepsilon' = 0.070$ cm^{-1} for pure water and 0.30 cm^{-1} for water exiting a waste water treatment plant. What distance corresponds to a decrease in I of 10% from its incident value for a) pure water, and b) waste water?

Using the empirical law:

$$\ln\left[\frac{I(\lambda)}{I_0(\lambda)}\right] = -\varepsilon'(\lambda)\ell$$

and solving for ℓ yields:

$$\ell = -\frac{\ln\left[\frac{I(\lambda)}{I_0(\lambda)}\right]}{\varepsilon'(\lambda)}$$

For pure water: $\ell = -\frac{\ln[0.9]}{0.070\text{cm}^{-1}} = 1.5\boldsymbol{cm}$

For treatment plant water: $\ell = -\frac{\ln[0.9]}{0.30\text{cm}^{-1}} = 0.35\boldsymbol{cm}$

The distance the UV light has to pass through to decrease 10% in intensity is longer for pure water.

Chapter 19: Electronic Spectroscopy

P19.1) Calculate the wavelengths of the first three lines of the Lyman, Balmer, and Paschen series, and the series limit (the shortest wavelength) for each series.

To calculate the wavelengths in the Lyman, Balmer, and Paschen series we use:

$$\tilde{\nu} = E_H\left(\frac{1}{n_i^2} - \frac{1}{n_f^2}\right)$$

And with $\lambda = 1/\tilde{\nu}$, and converting cm-1 to nm we obtain:

$$\lambda = \frac{1}{R_H\left(\frac{1}{n_i^2} - \frac{1}{n_f^2}\right)} \frac{1\,\text{nm}}{10^{-7}\,\text{cm}}$$

wave-length	$n_f = 2$	$n_f = 3$	$n_f = 4$	$n_f = 5$	$n_f = 6$	$n_f = 7$	$n_f = \infty$
Lymann $n_i = 1$	121.568	102.573	97.2548	94.9754	93.7814	93.0758	91.1763
Balmer $n_i = 2$	xxxx	656.47	486.274	434.173	410.294	397.124	364.705
Paschen $n_i = 3$	xxxx	xxxx	1875.63	1282.17	1094.12	1005.22	820.587

Chapter 20: Nuclear Magnetic Resonance Spectroscopy

P20.1) For a fixed frequency of the radio frequency field, ^{1}H, ^{13}C, and ^{31}P will be in resonance at different values of the static magnetic field. Calculate the value of B_0 for these nuclei to be in resonance if the radio-frequency field has a frequency of 250 MHz.

To obtain the field strength necessary to result in a resonance frequency of 250 MHz for ^{1}H, ^{31}P, and ^{13}C nuclei we solve:

$$B_0 = \frac{\nu_0 \, 2\pi}{\gamma}$$

for ν_0, and obtain:

$$\nu_0(^{1}H) = \frac{B_0 \, 2\pi}{\gamma(^{1}H)} = \frac{(250\times10^{6}\ s^{-1})2\pi}{(26.75\ rad\ T^{-1}\ s^{-1})} = \underline{5.87\ T}$$

$$\nu_0(^{31}P) = \frac{B_0 \, 2\pi}{\gamma(^{31}P)} = \frac{(250\times10^{6}\ s^{-1})2\pi}{(10.84\ rad\ T^{-1}\ s^{-1})} = \underline{14.49\ T}$$

$$\nu_0(^{13}C) = \frac{B_0 \, 2\pi}{\gamma(^{13}C)} = \frac{(250\times10^{6}\ s^{-1})2\pi}{(6.73\ rad\ T^{-1}\ s^{-1})} = \underline{23.34\ T}$$

P20.2) Using the information in Table 20.1, calculate the three Zeeman energies for a deuteron (^{2}H) in a magnetic field of 5.5 T. Calculate ΔE and the deuterium Larmor frequency in this field.

There are three energy levels for deuterium with I = 1 (2I+1). Using:

$$E = -\gamma \hbar \, m_z \, B_0,$$

we obtain the energies:

$$\underline{E(1) = -(4.11\times10^{7}\ rad\ T^{-1}\ s^{-1})\times(1.05457\times10^{-34}\ J\ s\ rad^{-1})\times(5.5\ T) = -2.3839\times10^{-26}\ J}$$

$$\underline{E(0) = 0 J}$$

$$\underline{E(-1) = (4.11\times10^{7}\ rad\ T^{-1}\ s^{-1})\times(1.05457\times10^{-34}\ J\ s\ rad^{-1})\times(5.5\ T) = 2.3839\times10^{-26}\ J}$$

P20.3) A 250-MHz ^{1}H spectrum of a compound shows two peaks. The frequency of one peak is 510 Hz higher than that of the reference compound (tetramethylsilane), and the second peak is at a frequency 280 Hz lower than that of the reference compound. What chemical shift should be assigned to these two peaks?

At this field strength, 250 Hz correspond to 1 ppm. Therefore:

$$\delta(280\text{ Hz}) = \frac{(1\text{ ppm})\times(280\text{ Hz})}{(250\text{ Hz})} = \underline{1.12\text{ ppm}}$$

$$\delta(-510\text{ Hz}) = \frac{(1\text{ ppm})\times(-510\text{ Hz})}{(250\text{ Hz})} = \underline{-2.04\text{ ppm}}$$

P20.7) In liquids, the amplitude of the spin echo is limited by irreversible dephasing that results when a molecule diffuses through an inhomogeneous magnet field during the duration of time between the application of the $\pi/2$ pulse and the echo time 2τ. In the presence of molecular diffusion through a magnetic field gradient ΔB, the proton spin echo amplitude is given by

$$I(2\tau) = I(0)\exp\left[-\left(\frac{2\tau}{T_2}\right) - \frac{2\gamma^2\Delta B^2 D\tau^3}{3}\right]$$

where D is the coefficient of diffusion. Because of the τ^3 dependence, molecular diffusion strongly affects measurements of T_2. Suppose $T_2 = 2$ s, $D = 1.00\times10^{-9}$ m^2 s^{-1}, and $\Delta B = 0.10$ T m^{-1}. To what fraction will the proton spin echo intensity be reduced at $\tau = 0.01$ s?

Solving $I(2\tau) = I(0)\exp\left[-\left(\frac{2\tau}{T_2}\right) - \frac{2\gamma^2\Delta B^2 D\tau^3}{3}\right]$ for $\frac{I(2\tau)}{I(0)}$ gives:

$$\frac{I(2\tau)}{I(0)} = \text{Exp}\left[-\left(\frac{2\tau}{T_2}\right) - \left(\frac{2\gamma^2\Delta B^2 D\tau^3}{3}\right)\right]$$

With: $\Delta B = 0.10\frac{T}{m}$, $D = 1.00\times10^{-9}\frac{m^2}{s}$, $T_2 = 2s$, $\tau = 0.01s$, and $\gamma = 4.11\times10^7\frac{\text{rad}}{\text{T s}}$:

$$\frac{I(2\tau)}{I(0)} = \text{Exp}\left[-\left(\frac{2\times 0.01\text{s}}{2\text{s}}\right) - \left(\frac{2\left(4.11\times 10^{7}\,\frac{\text{rad}}{\text{Ts}}\right)^2\left(0.10\frac{\text{T}}{\text{m}}\right)^2 1.00\times 10^{-9}\,\frac{\text{m}^2}{\text{s}}(0.01\text{s})^3}{3}\right)\right]$$

$$\frac{I(2\tau)}{I(0)} = 0.239$$

The spin echo will be reduced by 23.9%.

P20.21) Assume two protons in a rigid macromolecule are separated by 3.00×10^{-10} m. Assume the molecule tumbles with a correlation time of 20 ns. Estimate the rate of cross relaxation between these two spins.

In this limit the cross relaxation rate is given by: $W_0 \approx q\tau_C$

With $q = \frac{\hbar^2}{10}\left(\frac{\mu_0}{4\pi}\right)^2 \frac{\gamma_I^2\gamma_S^2}{r_{IS}^6}$, with $\hbar = 1.05457\times 10^{-34}\,\frac{\text{m}^2\text{ kg}}{\text{s rad}}$, $\gamma = \gamma_H = 26.75\times 10^7\,\frac{\text{A s rad}}{\text{kg}}$,

and $\mu_0 = 4\pi\times 10^{-7}\,\frac{\text{kg m}}{\text{s}^2\text{ A}^2}$, one obtains:

$$W_0 \approx \frac{\left(1.05457\times 10^{-34}\,\frac{\text{m}^2\text{kg}}{\text{s rad}}\right)^2}{10}\left(\frac{4\pi\times 10^{-7}\,\frac{\text{kg m}}{\text{s}^2\text{ A}^2}}{4\pi}\right)^2 \frac{\left(26.75\times 10^7\,\frac{\text{A s rad}}{\text{kg}}\right)^4}{\left(3\times 10^{-10}\text{m}\right)^6} \times \frac{\left(20\times 10^{-9}\text{s}\right)}{(2\pi\text{ rad})}$$

$$W_0 \approx 1.245\times 10^{-12}\,\frac{\text{rad}}{\text{s}}$$

P20.27) Calculate the dipolar coupling constant between a ^{13}C spin and a ^{19}F spin separated by 5 Å. Give your answer in radians per second. Calculate the dipolar splitting of the ^{13}C line by the dipolar coupling to this ^{19}F spin if the angle between the internuclear vector and the magnetic field is $\theta = \pi/3$.

The dipolar coupling constant between a ^{13}C spin and a ^{19}F spin is given by:

$$D = \frac{\mu_0}{4\pi}\frac{\gamma_{^{13}C}\gamma_{^{19}F}}{r_{CF}^{\ 3}}\frac{rad}{s}$$

And with:

$$\mu_0 = 4\pi\times10^{-7}\frac{N\ C}{s^2} = \frac{H}{m} = \frac{kg\ \ m^2}{s^2\ \ A^2\ \ m} = \frac{kg\ \ m}{s^2\ \ A^2}$$

$$\hbar = 1.05457\times10^{-34}\frac{kg\ \ m^2}{s\ \ rad} = \frac{kg\ \ m}{s^2\ \ A^2}$$

$$\gamma_C = 67.283\times10^6\frac{A\ \ s\ \ rad}{kg}$$

$$\gamma_F = 251.81\times10^6\frac{A\ \ s\ \ rad}{kg}$$

$$D = 1\times10^{-7}\frac{m\,kg}{s^2A^2}\times\frac{6.7283\times10^7\frac{A\,s\,rad}{kg}\times2.5181\times10^8\frac{A\,s\,rad}{kg}\times1.05457\times10^{-34}\frac{m^2\ kg}{s\,rad}}{\left(5\times10^{-10}\,m\right)^3}$$

$$D = 1429.37\frac{rad}{s}$$

Therefore the splitting, $\Delta\omega$, at $\theta = \frac{\pi}{3}$ is:

$$\Delta\omega = \left|2\times\frac{D}{2}\left(3Cos^2[\theta]-1\right)\right|$$

$$\Delta\omega = \left|1429.37\frac{rad}{s}\left(3Cos^2\left[\frac{\pi}{3}\right]-1\right)\right| = \left|1429.37\frac{rad}{s}\left(-\frac{1}{4}\right)\right|$$

$$\Delta\omega = 357.3\frac{rad}{s} = 56.87\frac{1}{s}$$

P20.29) Consider a ^{13}C spin with principal values $\sigma_1 = -81$ ppm, $\sigma_2 = 4$ ppm, and $\sigma_3 =$ 78 ppm and with a Larmor frequency of 125 MHz.

a. Calculate the chemical shift anisotropy, the chemical shift asymmetry, and the isotropic chemical shift

b. Calculate the chemical shift if the magnetic field is oriented at $\theta = \pi/4$ and $\varphi = 0$. Give your answer in rad s^{-1}.

a) For σ_1 = -81ppm, σ_2 = 4 ppm, and σ_3=78 ppm, ω_0 = 125 MHz:

$$\sigma_{iso} = \frac{1}{3}(\sigma_1 + \sigma_2 + \sigma_3) = \frac{1}{3}(\text{-}81\text{ppm} + 4\text{ppm} + 78\text{ppm}) = \frac{1}{3}\text{ppm}$$

$$\sigma_{aniso} = \sigma_3 - \sigma_{iso} = -81\text{ppm} - \frac{1}{3}\text{ppm} = -81.33\text{ppm}$$

$$\eta = \frac{(\sigma_1 - \sigma_2)}{(\sigma_3 - \sigma_{iso})} = \frac{(4\,ppm - 78\text{ppm})}{(\text{-}81\text{ppm} - \frac{1}{3}\text{ppm})} = 0.91$$

$$\text{b)}\ \sigma = \sigma_{iso} + \sigma_{aniso} = \sigma_{iso} + \delta_{aniso}\left[\frac{1}{2}(3\text{Cos}^2[\theta] - 1) + \frac{1}{2}\eta\text{Sin}^2[\theta]\text{Cos}[\varphi]\right]$$

$$\sigma = (\sigma_{iso} + \sigma_{aniso}) = \sigma_{iso} + \delta_{aniso}\left[\frac{1}{2}\left(3\text{Cos}^2\left[\frac{\pi}{4}\right] - 1\right) + \frac{1}{2}\eta\text{Sin}^2\left[\frac{\pi}{4}\right]\text{Cos}[0]\right]$$

$$\sigma = \sigma_{iso} + \sigma_{aniso} = \sigma_{iso} + \delta_{aniso}(0.25 + 0.25\eta)$$

$$\sigma = (0.33\text{ ppm}) + (81.33\text{ ppm}) \times (0.25 + 0.25 \times 0.91)$$

$$\sigma = 39.17\text{ ppm}$$

At a 125 MHz ^{13}C frequency:

125 Hz = 1ppm

Therefore:

$$\nu_{shift} = \frac{(125\text{Hz}) \times (39.17\text{ppm})}{(1\text{ppm})}$$

$$\nu_{shift} = 4895.6\text{ Hz} = 30760.18\frac{\text{rad}}{\text{s}}$$

P20.30) Calculate the ratio of the initial slopes of the NOE buildup curve for two amide proton pairs on adjacent amino acids, if one pair is located in an α-helix of a protein, and the other pair is located in a β-sheet of the same protein. Assume the amide proton pairs cross relax as isolated spin pairs.

Average distance in α-helix, $r_{helix} = 2.8$ Å

Average distance in β-sheet, $r_{sheet} = 2.8$ Å

With the slope $\propto r^{-6}$:

$$\frac{\text{Slope}(\alpha-\text{helix})}{\text{Slope}(\beta-\text{sheet})} = \frac{\dfrac{1}{(r_{helix})^6}}{\dfrac{1}{(r_{sheet})^6}} = \frac{(r_{sheet})^6}{(r_{helix})^6} = \frac{\left(3.5\,\text{Å}\right)^6}{\left(2.8\,\text{Å}\right)^6}$$

$$\frac{\text{Slope}(\alpha-\text{helix})}{\text{Slope}(\beta-\text{sheet})} = 3.81$$

Chapter 21: The Structure of Biomolecules at the Nanometer Scale: X-Ray Diffraction and Atomic Force Microscopy

P21.2) Taking sharing between neighboring cells into account, how many lattice points are contained in unit cells of the following Bravais lattices?

a. tetragonal I

b. cubic F

c. orthorhombic C

a) tetragonal I: number of points $= 8 \times 1/8 + 1 \times 1 = 2$

b) cubic F: number of points $= 8 \times 1/8 + 6 \times 1/2 = 4$

c) orthorhombic C: number of points $= 8 \times 1/8 + 2 \times 1/2 = 2$

P21.5) Using the tetragonal lattice unit cell shown below, draw in the following planes:

a. 001 **b.** 101 **c.** 112 **d.** 111

a) 001

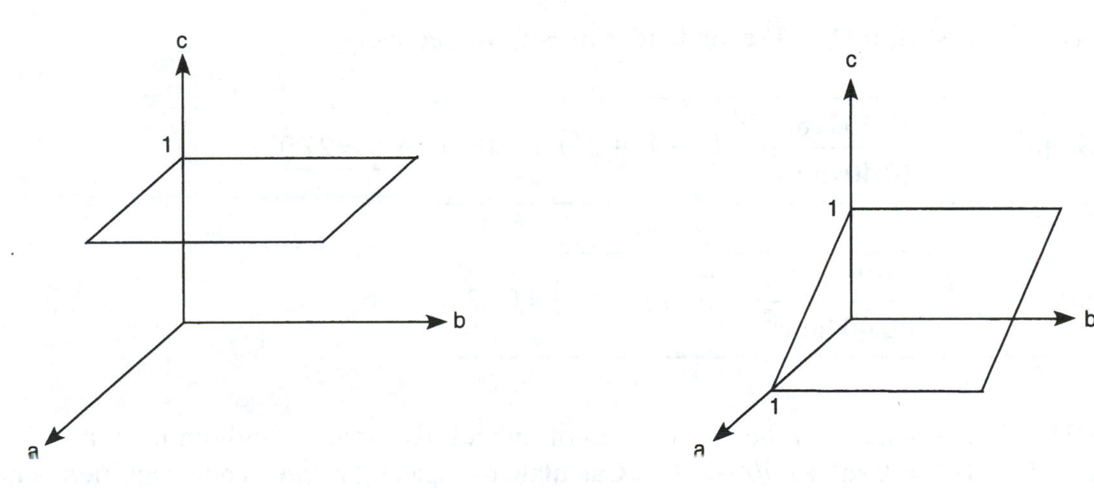

c) 112 d) 111

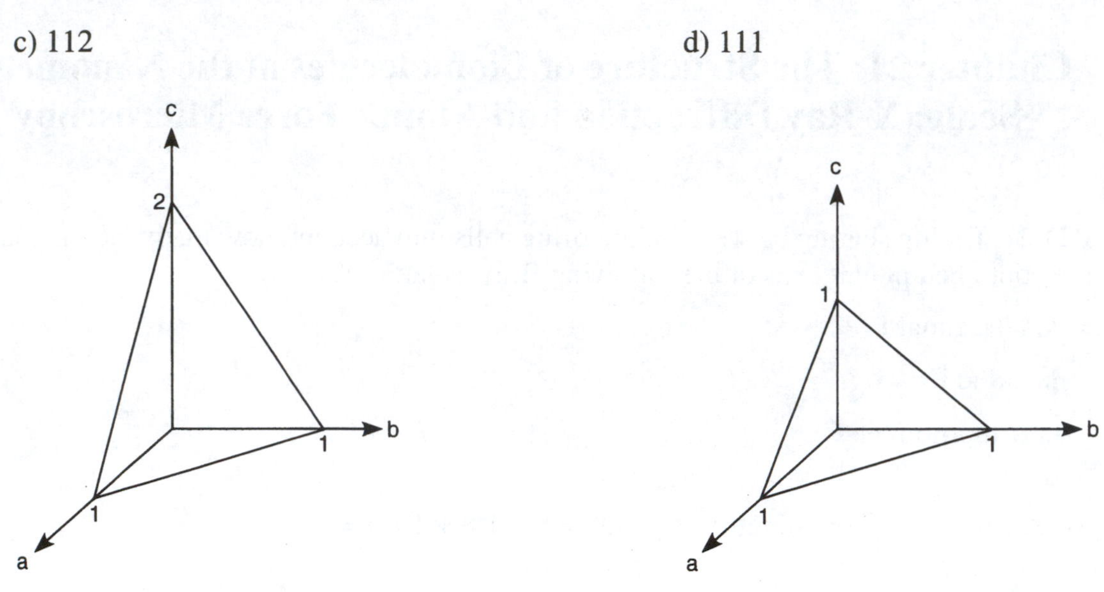

P21.13) Aluminum has a face-centered cubic structure with a = 0.404 nm. Calculate the angle θ at which diffraction is observed for (a) first-order and (b) second-order diffraction from the 112 planes. The X-ray wavelength is 0.15418 nm.

Using $a = 0.404\text{nm}$, $\lambda = 0.15418\text{nm}$, h = 1, k = 1, l = 2 in:

$$Sin[\theta] = \sqrt{\frac{n^2 \lambda^2}{4a^2}\left(h^2 + k^2 + l^2\right)}$$

gives for first (n = 1) and second order (n = 2), respectively:

$$\text{Sin}[\theta] = \sqrt{\frac{1^2 \times (0.15418\text{nm})^2}{4(0.404\text{nm})^2}\left(1^2 + 1^2 + 2^2\right)} = 0.4674 \rightarrow \underline{\theta = 27.9^\circ}$$

$$\text{Sin}[\theta] = \sqrt{\frac{2^2 \times (0.15418\text{nm})^2}{4(0.404\text{nm})^2}\left(1^2 + 1^2 + 2^2\right)} = 0.9348 \rightarrow \underline{\theta = 69.2^\circ}$$

P21.14) The spacing, d, between planes of indices hkl for an orthorhombic unit cell is given by $1/d^2 = h^2/a^2 + k^2/b^2 + l^2 / c^2$. Calculate the spacing between adjacent members of the (a) 011 and (b) 231 planes for unit cell lengths given by a = 512 pm, b = 498 pm, and c = 622 pm.

With $1/d^2 = h^2/a^2 + k^2/b^2 + l^2/c^2$ and a = 512 pm, b = 498 pm, and c = 622 pm

a) (011)

$$\frac{1}{d}=\sqrt{\frac{h^2}{a^2}+\frac{k^2}{b^2}+\frac{l^2}{c^2}}$$

$$\frac{1}{d}=\sqrt{\frac{0^2}{(512\text{pm})^2}+\frac{1^2}{(498\text{pm})^2}+\frac{1^2}{(622\text{pm})^2}}$$

$$\frac{1}{d}=\sqrt{6.6170\times10^{-6}\frac{1}{\text{pm}}} \text{ and } \underline{d=388.8\text{pm}}$$

b) (231)

$$\frac{1}{d}=\sqrt{\frac{2^2}{(512\text{pm})^2}+\frac{3^2}{(498\text{pm})^2}+\frac{1^2}{(622\text{pm})^2}}$$

$$\frac{1}{d}=\sqrt{5.413\times10^{-5}\frac{1}{\text{pm}}} \text{ and } \underline{d=135.9\text{pm}}$$

Chapter 22: The Boltzmann Distribution

P22.2)

a. Realizing that the most probable outcome from a series of N coin tosses is $N/2$ heads and $N/2$ tails, what is the expression for W_{max} corresponding to this outcome?

b. Given your answer for part (a), derive the following relationship between the weight for an outcome other than the most probable and W_{max}:

$$\log\left(\frac{W}{W_{\max}}\right) = -H\,\log\left(\frac{H}{N/2}\right) - T\log\left(\frac{T}{N/2}\right)$$

c. We can define the deviation of a given outcome from the most probable outcome using a "deviation index," $\alpha = (H - T)/N$. Show that the number of heads or tails can be expressed as $H = (N/2)\,(1 + \alpha)$ and $T = (N/2)(1 - \alpha)$.

d. Finally, demonstrate that $W/W_{max} = e^{-N\alpha^2}$.

a) W_{max} for the coin toss is given by:

$$W_{max} = \frac{N!}{(N/2)!\,(N/2)!} = \frac{N!}{((N/2)!)^2}$$

b) $$\ln\left(\frac{W}{W_{max}}\right) = \ln(W) - \ln(W_{max}) = \ln\left(\frac{N!}{H!\;T!}\right) - \ln\left(\frac{N!}{((N/2)!)^2}\right)$$

$$\ln\left(\frac{W}{W_{max}}\right) = \ln(N!) - \ln(H!) - \ln(T!) - \ln(N!) + 2\ln((N/2)!)$$

$$\ln\left(\frac{W}{W_{max}}\right) = -\ln(H!) - \ln(T!) + 2\ln((N/2)!)$$

$$\ln\left(\frac{W}{W_{max}}\right) = -H\ln(H) + H - T\ln(T) + T + N\ln(N/2) - N$$

$$\ln\left(\frac{W}{W_{max}}\right) = -H\ln(H) - T\ln(T) + N\ln(N/2) = -H\ln(H) - T\ln(T) + (H+T)\ln(N/2)$$

$$\ln\left(\frac{W}{W_{max}}\right) = -H\ln\left(\frac{H}{N/2}\right) - T\ln\left(\frac{T}{N/2}\right)$$

c) Using the definitions for the deviation indices:

$$H = \frac{N}{2}(1+\alpha) = \frac{N}{2}\left(1 + \frac{(H-T)}{N}\right) = \frac{N}{2} + \frac{(H-T)}{2} = \frac{(H+T)}{2} + \frac{(H-T)}{2} = H$$

$$H = \frac{N}{2}(1-\alpha) = \frac{N}{2}\left(1 - \frac{(H-T)}{N}\right) = \frac{N}{2} - \frac{(H-T)}{2} = \frac{(H+T)}{2} - \frac{(H-T)}{2} = T$$

d) Substituting the results from part c) into the expression derived in b) yields:

$$\ln\left(\frac{W}{W_{max}}\right) = -\frac{N}{2}(1+\alpha)\ln(1+\alpha) - \frac{N}{2}(1-\alpha)\ln(1-\alpha)$$

If $|\alpha| << 1$, then $\ln(1 \pm \alpha) = \pm\,\alpha$, therefore:

$$\ln\left(\frac{W}{W_{max}}\right) = -\frac{N}{2}(1+\alpha)\alpha - \frac{N}{2}(1-\alpha)\alpha = -N\alpha^2$$

$$\frac{W}{W_{max}} = \underline{e^{-N\alpha^2}}$$

P22.4) Determine the weight associated with the following card hands:

a. Having any five cards

b. Having five cards of the same suit (known as a "flush")

a) The statistical weight for drawing any five cards from a deck of 52 cards is:

$$W_{any\ five} = \frac{N!}{(N-5)!\ 5!} = \frac{52!}{(52-5)!\ 5!} = \underline{2.6 \times 10^6}$$

b) The statistical weight for drawing five cards of the same suit when there are 13 cards for each suit is:

$$W_{flush} = \frac{N!}{(N-5)!\ 5!} = \frac{13!}{(13-5)!\ 5!} = \underline{5148}$$

P22.7) A pair of standard dice are rolled. What is the probability of observing the following:

a. The sum of the dice is equal to 7.

b. The sum of the dice is equal to 9.

c. The sum of the dice is less than or equal to 7.

To calculate the probability for the events we need the weights for each outcome for rolling a die:

$$W_i = \frac{N!}{H_i!\,(N - H_i!)}$$

$$W_1 = \frac{1!}{1!\,(1 - 1!)} = W_2 = W_3 = W_4 = W_5 = W_6 = 1,$$

where i is the number on the die (1-6), and N is set to one toss. Then the probability for rolling each number is:

$$P_i = \frac{W_i}{\sum_i W_i} = \frac{1}{6}$$

Now we have to count what combinations of two dice produces in what outcome:

outcome	2	3	4	5	6	7	8	9	10	11	12
combinations	1 + 1	1 + 2 2 + 1	1 + 3 2 + 2 3 + 1	1 + 4 2 + 3 3 + 2 4 + 1	1 + 5 2 + 4 3 + 3 4 + 2 5 + 1	1 + 6 2 + 5 3 + 4 4 + 3 5 + 2 6 + 1	2 + 6 3 + 5 4 + 4 5 + 3 6 + 2	3 + 6 4 + 5 5 + 4 6 + 3	4 + 6 5 + 5 6 + 4	5 + 6 6 + 5	6 + 6
number of combinations	1	2	3	4	5	6	5	4	3	2	1

a) That means that the probability of rolling a 7 with two dice is:

$$P_7(2) = 2P_1P_6 + 2P_2P_5 + 2P_3P_4 = 6\left(\frac{1}{36}\right) = \text{combinations} \times \left(\frac{1}{36}\right) = \underline{\frac{1}{6}}$$

b) That means that the probability of rolling a 9 with two dice is:

$$P_9(2) = 2P_3P_6 + 2P_4P_5 = 4\left(\frac{1}{36}\right) = \underline{\frac{1}{9}}$$

c) That means that the probability of rolling a number less or equal to 7 with two dice is:

$$P_{\le 7}(2) = \sum_{i=1}^{7} P_i(2)$$
$$= (1P_1P_1) + (2P_2P_1) + (2P_1P_3 + P_2P_2) + (2P_1P_4 + 2P_2P_3)$$
$$+ (2P_1P_5 + P_3P_3 + 2P_2P_4) + (2P_1P_6 + 2P_2P_5 + 2P_3P_4)$$
$$= \left(\frac{1}{36}\right) + 2\left(\frac{1}{36}\right) + 3\left(\frac{1}{36}\right) + 4\left(\frac{1}{36}\right) + 5\left(\frac{1}{36}\right) + 6\left(\frac{1}{36}\right) = \left(\frac{21}{36}\right) = \underline{\frac{7}{12}}$$

P22.9) Four bases (A, C, T, and G) appear in DNA. Assume that the appearance of each base in a DNA sequence is random.

a. What is the probability of observing the sequence AAGACATGCA?

b. What is the probability of finding the sequence GGGGGAAAAA?

c. How do your answers to parts (a) and (b) change if the probability of observing A is twice that of the probabilities used in parts (a) and (b) of this question when the preceding base is G?

a) The probability for observing the sequence with the probability of choosing one of the four bases of ¼ for a sequence with 10 bases is:

$$P = (p_{base})^N = \left(\frac{1}{4}\right)^{10} = \underline{9.54\times10^{-7}}$$

b) With the same probabilities as in a) and the same sequence length, the answer is the same as in a):

$$P = (p_{base})^N = \left(\frac{1}{4}\right)^{10} = \underline{9.54\times10^{-7}}$$

c) We need to consider that every time there is a G, the probability of the following base being an A is (1/2), and of being a G,T or C is now (1/6).

For the first sequence, AAGACATGCA,

For the first three bases, the probability of each is (1/4)

NOW FOR THE fourth, the probability of it being an A is (1/2)

For the next 4 nucleotides, the probability of each is (1/4)

For the 9th base (after G) the probability of C is (1/6)

The final base has a probability of (1/4)

$$P = \left(\frac{1}{8}\right)^{8} \times \left(\frac{1}{2}\right)^{1} \times \left(\frac{1}{6}\right)^{1} = \underline{1.27\times10^{-7}}$$

For the second sequence, GGGGGAAAAA, everytime there is a G, the following nucleotide (if not A) has a probability of (1/6).

Therefore,

$$P = \left(\frac{1}{4}\right)^{5} \times \left(\frac{1}{2}\right)^{1} \times \left(\frac{1}{6}\right)^{4} = \underline{3.77\times10^{-7}}$$

P22.10) Imagine an experiment in which you flip a coin four times. Furthermore, the coin is balanced fairly such that the probability of landing heads or tails is equivalent. After tossing the coin 10 times, what is the probability of observing:

a. no heads?

b. two heads?

c. five heads?

d. eight heads?

To calculate the probability for the events we need to first calculate the weights:

$$W_{i,Heads} = \frac{N!}{H_i!\,(N - H_i!)},$$

where H_i is the number a head toss is observed, and N is the number of tosses. Then the probability is:

$$P_{i,Heads} = \frac{W_{i,Heads}}{\sum_i W_{i,Heads}}$$

a)

$$P_{0\,Heads} = \frac{\left(\frac{10!}{0!\,(10-0!)}\right)}{\left\{\left(\frac{10!}{0!\,(10-0!)}\right)+\left(\frac{10!}{1!\,(10-1!)}\right)+\left(\frac{10!}{2!\,(10-2!)}\right)+\ldots+\left(\frac{10!}{10!\,(10-10!)}\right)\right\}} = \underline{0.00977}$$

b) $$P_{2\,Heads} = \frac{\left(\frac{10!}{2!\,(10-2!)}\right)}{\left\{\left(\frac{10!}{0!\,(10-0!)}\right)+\left(\frac{10!}{1!\,(10-1!)}\right)+\left(\frac{10!}{2!\,(10-2!)}\right)+\ldots+\left(\frac{10!}{10!\,(10-10!)}\right)\right\}} = \underline{0.0439}$$

b) $$P_{5\,Heads} = \frac{\left(\frac{10!}{5!\,(10-5!)}\right)}{\left\{\left(\frac{10!}{0!\,(10-0!)}\right)+\left(\frac{10!}{1!\,(10-1!)}\right)+\left(\frac{10!}{2!\,(10-2!)}\right)+\ldots+\left(\frac{10!}{10!\,(10-10!)}\right)\right\}} = \underline{0.2461}$$

b) $$P_{8\,Heads} = \frac{\left(\frac{10!}{5!\,(10-5!)}\right)}{\left\{\left(\frac{10!}{0!\,(10-0!)}\right)+\left(\frac{10!}{1!\,(10-1!)}\right)+\left(\frac{10!}{2!\,(10-2!)}\right)+\ldots+\left(\frac{10!}{10!\,(10-10!)}\right)\right\}} = \underline{0.0439}$$

P22.14) Barometric pressure can be understood using the Boltzmann distribution. The potential energy associated with being a given height above the Earth's surface is *mgh,* where *m* is the mass of the particle of interest, *g* is the acceleration due to gravity, and *h* is height. Using this definition of the potential energy, derive the following expression for pressure: $P = P_0 e^{-mgh/kT}$. Assuming that the temperature remains at 298 K, what would you expect the relative pressures of N_2 and O_2 to be at the tropopause, the boundary between the troposphere and stratosphere roughly 11 km above the Earth's surface? At the Earth's surface, the composition of air is roughly 78% N_2, 21% O_2, and the remaining 1% is other gases.

Starting with the ideal gas law:

$$n = \frac{pV}{RT}$$

$$m = \frac{pVM}{RT}$$

$$\rho = \frac{m}{V} = \frac{pM}{RT} = \frac{pm}{kT}$$

Assuming only hydrostatic pressure, which decreases with height:

$$dp = -\rho g \, dh$$

We obtain:

$$dp = -\frac{p\,mg}{kT} dh$$

$$\frac{dp}{p} = -\frac{mg}{kT} dh$$

And integrating yields:

$$\int_{p_0}^{p} \frac{1}{p} dp = -\frac{mg}{kT} \int_0^{h_2} dh$$

$$\ln\left(\frac{p}{p_0}\right) = -\frac{mg}{kT} h$$

$$p = p_0 \operatorname{Exp}\left[-\frac{mg}{kT} h\right]$$

P22.16) Consider the following energy-level diagrams, modified from Problem P22.15 by the addition of another excited state with energy of 600 cm^{-1}:

a. At what temperature will the probability of occupying the second energy level be 0.15 for the states depicted in part (a) of the figure?

b. Perform the corresponding calculation for the states depicted in part (b) of the figure.

(*Hint:* You may find this problem easier to solve numerically using a spreadsheet program such as Excel.)

In general for the probability of occupying energy level n:

$$P_n = \frac{g_n \text{Exp}[-\beta\varepsilon_n]}{q} = \frac{g_n \text{Exp}[-\beta\varepsilon_n]}{\sum_n g_n \text{Exp}[-\beta\varepsilon_n]} = \frac{g_n \text{Exp}\left[-\frac{\varepsilon_n}{kT}\right]}{\sum_n g_n \text{Exp}\left[-\frac{\varepsilon_n}{kT}\right]}$$

a) The probability for occupying energy level 2 with no degeneracies in the system is:

$$P_n(T) = \frac{1\times \text{Exp}\left[-\frac{(30000\text{m}^{-1})\times(6.626\times10^{-34}\,\text{J s})\times(2.998\times10^{8}\,\text{m s}^{-1})}{(1.38\times10^{-23}\,\text{J K}^{-1})\times(\text{T/K})}\right]}{\sum_{n=1}^{3} g_n \text{Exp}\left[-\frac{\varepsilon_n}{k\,T}\right]}$$

Making a table to find the temperature gives a $P_n = 0.15$:

P_n	0.1470	0.1479	0.1487	0.1495	0.1503	0.1511	0.1519	0.1527	0.1535	0.1543
T	250	251	252	253	254	255	256	257	258	259

The temperature for which the probability of occupying energy level 2 is 254 K.

b) The probability for occupying energy level 2 is with that level being doubly degenerate is:

$$P_n(T) = \frac{2\times \text{Exp}\left[-\frac{(30000\text{m}^{-1})\times(6.626\times10^{-34}\,\text{J s})\times(2.998\times10^{8}\,\text{m s}^{-1})}{(1.38\times10^{-23}\,\text{J K}^{-1})\times(\text{T/K})}\right]}{\sum_{n=1}^{3} g_n \text{Exp}\left[-\frac{\varepsilon_n}{k\,T}\right]}$$

Making a table to find the temperature gives a $P_n = 0.15$:

P_n	0.145	0.146	0.148	0.149	0.151	0.153	0.154	0.156	0.158	0.160
T	275	276	277	278	279	280	281	282	283	284

The temperature for which the probability of occupying energy level 2 is about 279 K.

P22.21) The ^{13}C nucleus is a spin 1/2 particle as is a proton. However, the energy splitting for a given field strength is roughly 1/4 of that for a proton. Using a 1.45-T magnet as in Example Problem 22.7, what is the ratio of populations in the excited and ground spin states for ^{13}C at 298 K?

The population difference between the ground and excited state for the ^{13}C nucleus is given by:

$$\frac{a_{\frac{1}{2}}}{a_{-\frac{1}{2}}} = \mathrm{Exp}[-\beta\Delta E] = \mathrm{Exp}\left[-\frac{(2.82\times10^{-26}\,\mathrm{J\,T^{-1}})\times(1.45\,\mathrm{T})}{4\times(1.38\times10^{-23}\,\mathrm{J\,K^{-1}})\times(298\,\mathrm{K})}\right] = \underline{0.999998}$$

P22.23) The vibrational frequency of I_2 is 208 cm^{-1}. At what temperature will the population in the first excited state be half that of the ground state?

The temperature at which the population of the ground state is twice that of the first excited state is given by:

$$\frac{1}{2} = \mathrm{Exp}[-\beta\Delta E] = \mathrm{Exp}\left[-\frac{(E_1 - E_1)}{k\,T}\right]$$

Solving for T yields:

$$T = \frac{(E_1 - E_1)}{\ln\left(\frac{1}{2}\right)k} = -\frac{(4.13\times10^{-21}\,\mathrm{J})}{\ln\left(\frac{1}{2}\right)\times(1.38\times10^{-23}\,\mathrm{J\,K^{-1}})} = \underline{431.5\,\mathrm{K}}$$

P22.27) The lowest two electronic energy levels of the molecule NO are illustrated here: Determine the probability of occupying one of the higher energy states at 100, 500, and 2000 K.

In general for the probability of occupying energy level n:

$$P_n = \frac{g_n \mathrm{Exp}[-\beta\varepsilon_n]}{q} = \frac{g_n \mathrm{Exp}[-\beta\varepsilon_n]}{\sum_n g_n \mathrm{Exp}[-\beta\varepsilon_n]} = \frac{g_n \mathrm{Exp}\left[-\frac{\varepsilon_n}{kT}\right]}{\sum_n g_n \mathrm{Exp}\left[-\frac{\varepsilon_n}{kT}\right]}$$

The probability for occupying energy level 2, which is doubly degenerate, and considering the double degeneracy of level 1, can be calculated as follows:

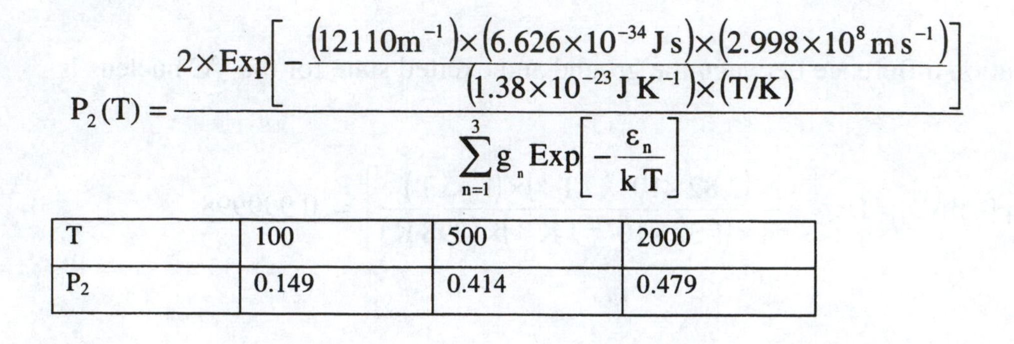

$$P_2(T) = \frac{2\times \mathrm{Exp}\left[-\dfrac{(12110\,\mathrm{m}^{-1})\times(6.626\times10^{-34}\,\mathrm{J\,s})\times(2.998\times10^{8}\,\mathrm{m\,s}^{-1})}{(1.38\times10^{-23}\,\mathrm{J\,K}^{-1})\times(\mathrm{T/K})}\right]}{\sum_{n=1}^{3} g_n\,\mathrm{Exp}\left[-\dfrac{\varepsilon_n}{k\,T}\right]}$$

T	100	500	2000
P_2	0.149	0.414	0.479

Chapter 23: Statistical Thermodynamics

P23.2) Evaluate the translational partition function for Ar confined to a volume of 1000 cm^3 at 298 K. At what temperature will the translational partition function of Ne be identical to that of Ar at 298 K confined to the same volume?

The translational partition function, q_T, of argon can be calculated using:

$$q_T(Ar) = (2\pi\ m_{Ar}\ k\,T)^{3/2} \frac{V}{h^3} =$$

$$\left(2\pi\left(6.63401\times10^{-26}\ kg\right)\times\left(1.3807\times10^{-23}\ J\,K^{-1}\right)\times\left(298\ K\right)\right)^{3/2} \frac{\left(0.001\ m^3\right)}{\left(6.626\times10^{-34}\ J\,s\right)^3} = \underline{2.44\times10^{29}}$$

The temperature at which Ne would have the same translational partition function is obtained by solving the above equation for q_T for T and using the atomic mass of Ne:

$$T = \sqrt[3]{\left(\frac{q_T\ h^3}{\left(V\ 2^{3/2}\ \pi^{3/2}\ m_{Ne}^{\ 3/2}\ k^{3/2}\right)}\right)} =$$

$$\sqrt[3]{\left(\frac{\left(2.44\times10^{29}\right)\times\left(6.63401\times10^{-26}\ J\,kg\right)^3}{\left(\left(0.001\ m^3\right)2^{3/2}\ \pi^{3/2}\left(3.35105\times10^{-26}\ J\,kg\right)^{3/2}\left(1.3807\times10^{-23}\ J\,K^{-1}\right)^{3/2}\right)}\right)} = \underline{589.9\ K}$$

P23.5) For N_2 at 77.3 K, 1 atm, in a 1-cm^3 container, calculate the translational partition function and ratio of this partition function to the number of N_2 molecules present under these conditions.

The translational partition function, q_T, of nitrogen is:

$$q_T(N_2) = \left(2\pi\ m_{N_2}\ k\,T\right)^{3/2} \frac{V}{h^3} =$$

$$\left(2\pi\left(\ 4.65294\times10^{-26}\ kg\right)\times\left(1.3807\times10^{-23}\ J\,K^{-1}\right)\times\left(77.3\ K\right)\right)^{3/2} \frac{\left(10^{-6}\ m^3\right)}{\left(6.626\times10^{-34}\ J\,s\right)} = \underline{1.89\times10^{25}}$$

We can use the ideal gas law to determine the number of N_2 molecules in the container:

$$nmol(N_2) = \frac{p\,V\,N_A}{R\,T} = \frac{\left(101325\ Pa\right)\times\left(10^{-6}\ m^3\right)\times\left(6.022\times10^{23}\ mol^{-1}\right)}{\left(8.314472\ J\,K^{-1}\ mol^{-1}\right)\times\left(77.3\ K\right)} = 9.49\times10^{19}$$

The ratio of the translational partition function and the number of N_2 molecules is then:

$$ratio = \frac{q_T(N_2)}{nmol(N_2)} = \frac{\left(1.89\times10^{25}\right)}{\left(9.49\times10^{19}\right)} = \underline{2.00\times10^{5}}$$

P23.9) Calculate the rotational partition function for SO_2 at 298 K where $B_A = 2.03\ cm^{-1}$, $B_B = 0.344\ cm^{-1}$, and $B_C = 0.293\ cm^{-1}$.

The rotational partition function, q_R, of SO_2 can be obtained from:

$$q_R(SO_2) = \frac{\sqrt{\pi}}{\sigma}\left(\frac{1}{(\beta h c B_A)}\right)^{1/2}\left(\frac{1}{(\beta h c B_B)}\right)^{1/2}\left(\frac{1}{(\beta h c B_C)}\right)^{1/2}$$

With $\beta = 1/(kT)$ and the symmetry number $\sigma = 2$ (two equivalent positions for the rotation of the SO_2 molecule about the axis in the O-S-O plane, bisecting the O-S-O angle) we get:

$$q_R(SO_2) = \frac{\sqrt{\pi}}{\sigma}\left(\frac{kT}{(h c B_A)}\right)^{1/2}\left(\frac{kT}{(\beta h c B_A)}\right)^{1/2}\left(\frac{kT}{(\beta h c B_A)}\right)^{1/2}$$

$$= \frac{\sqrt{\pi}}{2}\left(\frac{(1.3807\times10^{-23}\ J\ K^{-1})\times(298\ K)}{(6.626\times10^{-34}\ J\ s)\times(2.998\times10^{8}\ m\ s^{-1})\times(203\ m^{-1})}\right)^{1/2}$$

$$\left(\frac{(1.3807\times10^{-23}\ J\ K^{-1})\times(298\ K)}{(6.626\times10^{-34}\ J\ s)\times(2.998\times10^{8}\ m\ s^{-1})\times(34.4\ m^{-1})}\right)^{1/2}$$

$$\left(\frac{(1.3807\times10^{-23}\ J\ K^{-1})\times(298\ K)}{(6.626\times10^{-34}\ J\ s)\times(2.998\times10^{8}\ m\ s^{-1})\times(29.3\ m^{-1})}\right)^{1/2} = \underline{5840.3}$$

P23.12)

a. Calculate the percent population of the first 10 rotational energy levels for HBr ($B = 8.46\ cm^{-1}$) at 298 K.

b. Repeat this calculation for HF assuming that the bond length of this molecule is identical to that of HBr.

a) The percent populations of the rotational energy levels are calculated from:

$$P_J = \frac{(2J+1)\mathrm{Exp}[-\beta h c B J(J+1)]}{q_R}, \text{ with } q_R = \frac{1}{(\sigma \beta h c B)} = \frac{(kT)}{(\sigma h c B)}$$

Therefore:

$$P_J(\text{HBr}) = \frac{(2J+1)\ (kT)\ \text{Exp}\left[-\beta\, h\, c\, B_{HBr}\ J\ (J+1)\right]}{(\sigma\, h\, c\, B_{HBr})}$$

$$= \frac{(2J+1)\ \left(1.3807\times10^{-23}\ \text{J K}^{-1}\right)\times(298\ \text{K})\ \text{Exp}\left[-\frac{\left(6.626\times10^{-34}\ \text{J s}\right)\times\left(2.998\times10^{8}\ \text{m s}^{-1}\right)\times\left(846\ \text{m}^{-1}\right)}{\left(1.3807\times10^{-23}\ \text{J K}^{-1}\right)\times(298\ \text{K})}\ J\ (J+1)\right]}{1\times\left(6.626\times10^{-34}\ \text{J s}\right)\times\left(2.998\times10^{8}\ \text{m s}^{-1}\right)\times\left(846\ \text{m}^{-1}\right)}$$

J	0	1	2	3	4	5	6	7	8	9
P_J	0.0408	0.1129	0.1598	0.1751	0.1624	0.1319	0.0955	0.0622	0.0367	0.0197

b) To do the same calculation as in a) we first need to extract the bond length of HBr. To do that we use:

$$B = \frac{h}{8\pi^2 c I} = \frac{h}{8\pi^2 c \mu r^2} = \frac{h}{8\pi^2 c\left(\frac{m_1 m_2}{m_1 + m_2}\right) r^2},$$

and solve for r:

$$r(\text{HBr}) = \sqrt{\frac{h}{8\pi^2 c\left(\frac{m_H m_{Br}}{m_H + m_{Be}}\right) B_{HBr}}}$$

$$= \sqrt{\frac{\left(6.626\times10^{-34}\ \text{J s}\right)}{8\pi^2\left(2.998\times10^{8}\ \text{m s}^{-1}\right)\times\left(\frac{\left(1.67718\times10^{-27}\ \text{kg}\right)\times\left(1.3268\times10^{-27}\ \text{kg}\right)}{\left(1.67718\times10^{-27}\ \text{kg}\right)+\left(1.3268\times10^{-27}\ \text{kg}\right)}\right)\left(846\ \text{m}^{-1}\right)}} = 1.41341\times10^{-10}\ \text{m}$$

Now, we can calculate B for HF:

$$B = \frac{h}{8\pi^2 c \left(\dfrac{m_H\, m_F}{m_H + m_F}\right) r_{HBr}^{\;2}}$$

$$= \frac{\left(6.626\times10^{-34}\ \mathrm{J\,s}\right)}{8\pi^2\left(2.998\times10^{8}\ \mathrm{m\,s^{-1}}\right)\times\left(\dfrac{\left(1.67718\times10^{-27}\ \mathrm{kg}\right)\times\left(3.1551\times10^{-27}\ \mathrm{kg}\right)}{\left(1.67718\times10^{-27}\ \mathrm{kg}\right)+\left(3.1551\times10^{-27}\ \mathrm{kg}\right)}\right)\left(1.41341\times10^{-10}\ \mathrm{m}\right)^2}$$

$$= 1.41341\times10^{-10}\ \mathrm{m}$$

$$\sqrt{\frac{\left(6.626\times10^{-34}\ \mathrm{J\,s}\right)}{8\pi^2\left(2.998\times10^{8}\ \mathrm{m\,s^{-1}}\right)\times\left(\dfrac{\left(1.67718\times10^{-27}\ \mathrm{kg}\right)\times\left(1.3268\times10^{-27}\ \mathrm{kg}\right)}{\left(1.67718\times10^{-27}\ \mathrm{kg}\right)+\left(1.3268\times10^{-27}\ \mathrm{kg}\right)}\right)\left(846\ \mathrm{m^{-1}}\right)}} = 879.85\ \mathrm{m^{-1}}$$

Then finally, the P_J values for HF are:

$$P_J = \frac{(2J+1)(kT)\mathrm{Exp}\left[-\beta h c B_{HF} J(J+1)\right]}{(\sigma h c B_{HF})}$$

$$= \frac{(2J+1)\left(1.3807\times10^{-23}\ \mathrm{J\,K^{-1}}\right)\times(298\ \mathrm{K})\mathrm{Exp}\left[-\dfrac{\left(1.3807\times10^{-23}\ \mathrm{J\,K^{-1}}\right)\times(298\ \mathrm{K})}{\left(6.626\times10^{-34}\ \mathrm{J\,s}\right)\times\left(2.998\times10^{8}\ \mathrm{m\,s^{-1}}\right)\times\left(879.85\ \mathrm{m^{-1}}\right)}\,J(J\right.}{1\times\left(6.626\times10^{-34}\ \mathrm{J\,s}\right)\times\left(2.998\times10^{8}\ \mathrm{m\,s^{-1}}\right)\times\left(879.85\ \mathrm{m^{-1}}\right)}$$

J	0	1	2	3	4	5	6	7	8	9
P_J	0.0425	0.1171	0.1646	0.1786	0.1635	0.1307	0.0927	0.0590	0.0339	0.0176

P23.16) Evaluate the vibrational partition function for H_2O at 2000 K where the vibrational frequencies are 1615, 3694, and 3802 cm^{-1}.

The vibrational partition function of water, a nonlinear polyatomic molecule, is given by:

$$q_V = \prod_{i=1}^{3N-6}\left(q_{V,i}\right) = \prod_{i=1}^{3N-6}\frac{1}{1-\mathrm{Exp}\left[-\beta h c \tilde{\nu}\right]}$$

With N = 3, we obtain:

$$q_V = \frac{1}{1\text{-Exp}\left[-\frac{(6.626\times10^{-34}\text{ J s})\times(2.998\times10^{8}\text{ m s}^{-1})\times(161500\text{ m}^{-1})}{(1.3807\times10^{-23}\text{ J K}^{-1})\times(2000\text{ K})}\right]}$$

$$\times\frac{1}{1\text{-Exp}\left[-\frac{(6.626\times10^{-34}\text{ J s})\times(2.998\times10^{8}\text{ m s}^{-1})\times(369400\text{ m}^{-1})}{(1.3807\times10^{-23}\text{ J K}^{-1})\times(2000\text{ K})}\right]}$$

$$\times\frac{1}{1\text{-Exp}\left[-\frac{(6.626\times10^{-34}\text{ J s})\times(2.998\times10^{8}\text{ m s}^{-1})\times(380200\text{ m}^{-1})}{(1.3807\times10^{-23}\text{ J K}^{-1})\times(2000\text{ K})}\right]} = \underline{1.67}$$

P23.18) Evaluate the vibrational partition function for NH_3 at 1000 K for which the vibrational frequencies are 950, 1627.5 (doubly degenerate), 3335, and 3414 cm^{-1} (doubly degenerate). Are there any modes that you can disregard in this calculation? Why or why not?

The vibrational partition function of NH_3, a non-linear polyatomic molecule, is given by:

$$q_V = \prod_{i=1}^{3N-6}(q_{V,i}) = \prod_{i=1}^{3N-6}\frac{1}{1-\text{Exp}[-\beta h c \tilde{\nu}]}$$

With N = 4, we obtain:

$$q_V = \prod_{i}^{6}\frac{1}{1\text{-Exp}\left[-\frac{(6.626\times10^{-34}\text{ J s})\times(2.998\times10^{8}\text{ m s}^{-1})\times(\nu_i\text{ m}^{-1})}{(1.3807\times10^{-23}\text{ J K}^{-1})\times(1000\text{ K})}\right]} = \underline{1.68}$$

Alternatively, we could have used:

$$q_V = \prod_{i=1}^{n'}\left(q_{V,i}^{\,g_i}\right),$$

where n' is the number of unique energy levels, and each partition function in the product is raised to the degeneracy level. In any case, the degeneracy has to be considered and no energy can be omitted in the calculation.

P23.22) Consider a particle free to translate in one dimension. The classical Hamiltonian is $H = p^2/2m$.

a. Determine $q_{classical}$ for this system. To what quantum system should you compare it in order to determine the equivalence of the classical and quantum statistical mechanical treatments?

b. Derive $q_{classical}$ for a system with translational motion in three dimensions for which

$$H = \left(p_x^{\ 2} + p_x^{\ 2} + p_x^{\ 2}\right)/\ 2m$$

The classical partition function can be calculated as:

$$q_{classical} = \frac{1}{h^{3N}} \int \mathrm{Exp}[-\beta H] dp^{3N} dx^{3N}$$

a) The classical partition function for the system is:

$$q_{classical} = \frac{1}{h} \int_{-\infty}^{\infty} \mathrm{Exp}\left[-\frac{p^2}{2\,m\,k\,T}\right] dp$$

With:

$$\int_{-\infty}^{\infty} \mathrm{Exp}\left[-c\,x^2\right] dx = \sqrt{\frac{\pi}{c}}$$

We get:

$$q_{classical} = \frac{1}{h} \int_{-\infty}^{\infty} \mathrm{Exp}\left[-\frac{p^2}{2\,m\,k\,T}\right] dp \ = \underline{\frac{1}{h}\sqrt{2\,\pi\,m\,k\,T}}$$

This should be compared to the one-dimensional particle in a box quantum mechanical system.

b) The classical partition function for this system is:

$$q_{classical} = \frac{1}{h^3} \int_{-\infty}^{\infty} \mathrm{Exp}\left[-\frac{1}{k\,T}\left(\frac{p_x^{\ 2}}{2\,m} + \frac{p_y^{\ 2}}{2\,m} + \frac{p_z^{\ 2}}{2\,m}\right)\right] dp_x\, dp_y\, dp_z$$

$$= \frac{1}{h^3} \int_{-\infty}^{\infty} \mathrm{Exp}\left[-\frac{p_x^{\ 2}}{2\,m\,k\,T}\right] dp_x\ \mathrm{Exp}\left[-\frac{p_y^{\ 2}}{2\,m\,k\,T}\right] dp_y\ \mathrm{Exp}\left[-\frac{p_z^{\ 2}}{2\,m\,k\,T}\right] dp_z = \underline{\frac{1}{h^3}(2\,\pi\,m\,k\,T)^{3/2}}$$

P23.24) Determine the total molecular partition function for I_2, confined to a volume of $1000\ cm^3$ at 298 K. Other information you will find useful: $B = 0.0374\ cm^{-1}$, $\tilde{\nu} = 208\ cm^{-1}$, and the ground electronic state is nondegenerate.

The total molecular partition function is given by the product of the individual partition functions:

$$q_{total} = q_T\, q_R\, q_V\, q_E$$

$$q_{total} = \left\{(2\pi\, m\, k\, T)^{3/2}\frac{V}{h^3}\right\}\left\{\frac{k\,T}{(\sigma\, h\, c\, B_A)}\right\}\left\{\frac{1}{(1-\mathrm{Exp}[-\beta\, h\, c\, \tilde{\nu}])}\right\}\left\{\sum_n g_n\ \mathrm{Exp}[-\beta\, E_n]\right\}$$

With the symmetry number $\sigma = 2$, and the electronic partition function in first approximation $q_E \approx g_0 \approx 1$ we get:

$$q_{total} = \left\{\left(2\pi\left(4.21455\times10^{-25}\ \mathrm{kg}\right)\times\left(1.3807\times10^{-23}\ \mathrm{J\,K^{-1}}\right)\times(298\ \mathrm{K})\right)^{3/2}\frac{\left(0.001\ \mathrm{m^3}\right)}{\left(6.626\times10^{-34}\ \mathrm{J\,s}\right)^3}\right\}$$

$$\times\left\{\frac{\left(1.3807\times10^{-23}\ \mathrm{J\,K^{-1}}\right)\times(298\ \mathrm{K})}{2\times\left(6.626\times10^{-34}\ \mathrm{J\,s}\right)\times\left(2.998\times10^{8}\ \mathrm{m\,s^{-1}}\right)\times\left(3.74\ \mathrm{m^{-1}}\right)}\right\}$$

$$\times\left\{\frac{1}{\left(1-\mathrm{Exp}\left[-\frac{\left(6.626\times10^{-34}\ \mathrm{J\,s}\right)\times\left(2.998\times10^{8}\ \mathrm{m\,s^{-1}}\right)\times\left(20800\ \mathrm{m^{-1}}\right)}{\left(1.3807\times10^{-23}\ \mathrm{J\,K^{-1}}\right)\times(298\ \mathrm{K})}\right]\right)}\right\}$$

$$\times\{1\} = \underline{1.71\times10^{34}}$$

P23.28) Consider an ensemble of units in which the first excited electronic state at energy ε_1 is m_1-fold degenerate, and the energy of the ground state is m_o-fold degenerate with energy ε_0.

a. Demonstrate that if $\varepsilon_0 = 0$, the expression for the electronic partition function is

$$q_E = m_o\left(1+\frac{m_1}{m_o}e^{-\varepsilon_1/kT}\right)$$

b. Determine the expression for the internal energy U of an ensemble of N such units. What is the limiting value of U as the temperature approaches zero and infinity?

a) The electronic partition function is given by:

$$q_{electronic} = \sum_n g_n\ \mathrm{Exp}[-\beta\, E_n]$$

with $E_0 = 0$:

$$q_{electronic} = m_0 + m_1\ \mathrm{Exp}[-\beta\,\varepsilon_1] = \underline{m_0\left(1+\frac{m_1}{m_0}\mathrm{Exp}[-\beta\,\varepsilon_1]\right)}$$

b) An expression for the internal energy can be derived as follows:

$$U=-\left(\frac{d\ln Q}{d\beta}\right)_V=-n\,N_A\left(\frac{d\ln q}{d\beta}\right)_V=-n\,N_A\frac{1}{q}\left(\frac{dq}{d\beta}\right)_V$$

Examining the derivative of q:

$$\left(\frac{dq}{d\beta}\right)_V=\frac{d}{d\beta}(m_0+m_1\,\text{Exp}[-\varepsilon_1\,\beta])_V=-\varepsilon_1\,m_1\,\text{Exp}[-\varepsilon_1\,\beta]=-\varepsilon_1\,(q_{electronic}-m_0)$$

Therefore:

$$U=-n\,N_A\frac{1}{q_{electronic}}(\varepsilon_1\,m_1\,\text{Exp}[-\varepsilon_1\,\beta])=-n\,N_A\frac{1}{q_{electronic}}(\varepsilon_1\,(q_{electronic}-m_0))$$

$$=-n\,N_A\frac{1}{q_{electronic}}(\varepsilon_1\,q_{electronic}-\varepsilon_1\,m_0)=-n\,N_A\left(\varepsilon_1-\varepsilon_1\frac{m_0}{q_{electronic}}\right)=-n\,N_A\,\varepsilon_1\left(1-\frac{m_0}{q_{electronic}}\right)$$

$$=-n\,N_A\,\varepsilon_1\left(1-\frac{m_0}{\left(m_0+m_1\,\text{Exp}\left[-\frac{\varepsilon_1}{k\,T}\right]\right)}\right)$$

$$U_{T\to\infty}=-n\,N_A\,\varepsilon_1\left(1-\frac{m_0}{(m_0+m_1)}\right)=-n\,N_A\,\varepsilon_1\left(\frac{m_1}{(m_0+m_1)}\right)$$

$$U_{T\to 0}=0$$

P23.31) Determine the vibrational contribution to C_V for HCl $\left(\tilde{\nu}=2886\text{ cm}^{-1}\right)$ over a temperature range from 500 to 5000 K in 500-K intervals and plot your result. At what temperature do you expect to reach the high-temperature limit for the vibrational contribution to C_V?

The total vibrational contribution to C_V for HCl is given by:

$$(C_V)_{vib,tot}=\sum_{k=1}^{3N-5}(C_V)_{vib,tot,k}=N\,k\,\beta^2(h\,c\,\tilde{\nu})^2\frac{\text{Exp}[-\beta\,h\,c\,\tilde{\nu}]}{(\text{Exp}[-\beta\,h\,c\,\tilde{\nu}]-1)^2}$$

$$(C_V)_{vib,tot}=\sum_{k=1}^{3N-5}(C_V)_{vib,tot,k}=\frac{N}{k\,T^2}(h\,c\,\tilde{\nu})^2\frac{\text{Exp}\left[-\frac{h\,c\,\tilde{\nu}}{k\,T}\right]}{\left(\text{Exp}\left[-\frac{h\,c\,\tilde{\nu}}{k\,T}\right]-1\right)^2}$$

$$(C_V)_{vib,tot} = \sum_{k=1}^{3N-5} (C_V)_{vib,tot,k} = \frac{(6.022\times10^{23}\ \text{mol}^{-1})}{(1.3807\times10^{-23}\ \text{J K}^{-1})\times(T)^2}$$

$$\times\left((6.626\times10^{-34}\ \text{J s})\times(2.998\times10^{8}\ \text{m s}^{-1})\times(2886\ \text{m}^{-1})\right)^2$$

$$\times\frac{\text{Exp}\left[-\frac{(6.626\times10^{-34}\ \text{J s})\times(2.998\times10^{8}\ \text{m s}^{-1})\times(2886\ \text{m}^{-1})}{(1.3807\times10^{-23}\ \text{J K}^{-1})\times(T)}\right]}{\left(\text{Exp}\left[-\frac{(6.626\times10^{-34}\ \text{J s})\times(2.998\times10^{8}\ \text{m s}^{-1})\times(2886\ \text{m}^{-1})}{(1.3807\times10^{-23}\ \text{J K}^{-1})\times(T)}\right]-1\right)^2}$$

Plotting $(C_V)_{vib,tot}$ as a function of T/K yields:

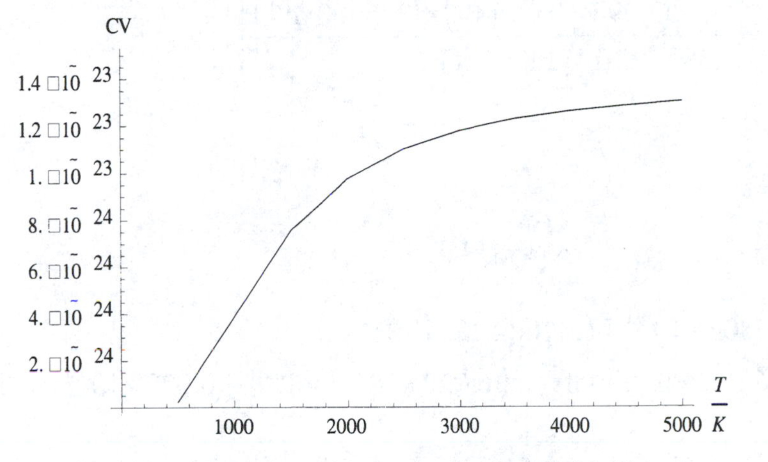

The high-temperature limit is reached at 5000 K.

P23.35) Determine the molar entropy of N_2 ($\tilde{\nu} = 2359\ \text{cm}^{-1}$ and $B = 2.00\ \text{cm}^{-1}$, $g_0 = 1$) and the entropy when $P = 1$ atm, but $T = 2500$ K.

The molar molecular entropy of N_2 is given by:

$$S_{molar} = \frac{3}{2}N_A k + N_A k \ln(q) - k(N_A \ln(N_A) - N_A)$$

We first calculate the total partition function at T = 2500:

$$q_{total} = \left\{ \frac{\left(2\pi\left(4.65294\times10^{-26}\ \text{kg}\right)\times\left(1.3807\times10^{-23}\ \text{J K}^{-1}\right)\times(2500\ \text{K})\right)^{3/2} \times \left(\frac{\left(8.314472\ \text{J mol}^{-1}\ \text{K}^{-1}\right)\times(2500\ \text{K})}{(101325\ \text{Pa})}\right)}{\left(6.626\times10^{-34}\ \text{J s}\right)^3} \right\}$$

$$\times\left\{\frac{\left(1.3807\times10^{-23}\ \text{J K}^{-1}\right)\times(2500\ \text{K})}{2\times\left(6.626\times10^{-34}\ \text{J s}\right)\times\left(2.998\times10^{8}\ \text{m s}^{-1}\right)\times\left(200\ \text{m}^{-1}\right)}\right\}$$

$$\times\left\{\frac{1}{\left(1-\text{Exp}\left[-\frac{\left(6.626\times10^{-34}\ \text{J s}\right)\times\left(2.998\times10^{8}\ \text{m s}^{-1}\right)\times\left(235900\ \text{m}^{-1}\right)}{\left(1.3807\times10^{-23}\ \text{J K}^{-1}\right)\times(\text{T})}\right]\right)}\right\}$$

$$\times\{1\} = 4.18113\times10^{35}$$

Then:

$$S_{molar} = \frac{3}{2}\left(6.022\times10^{23}\ \text{mol}^{-1}\right)\times\left(1.3807\times10^{-23}\ \text{J K}^{-1}\right)$$
$$+\left(6.022\times10^{23}\ \text{mol}^{-1}\right)\times\left(1.3807\times10^{-23}\ \text{J K}^{-1}\right)\ln\left(4.18113\times10^{35}\right)$$
$$-\left(1.3807\times10^{-23}\ \text{J K}^{-1}\right)\times\left(\left(6.022\times10^{23}\ \text{mol}^{-1}\right)\ln\left(\left(6.022\times10^{23}\ \text{mol}^{-1}\right)\right)-\left(6.022\times10^{23}\ \text{mol}^{-1}\right)\right)$$
$$= \underline{247.5\ \text{J mol}^{-1}\ \text{K}^{-1}}$$

P23.38) The molecule NO has a ground electronic level that is doubly degenerate, and a first excited level at 121.1 cm^{-1} that is also twofold degenerate. Determine the contribution of electronic degrees of freedom to the standard molar entropy of NO. Compare your result to $R\ \ln(4)$. What is the significance of this comparison?

The contribution of the degrees of freedom to the molar molecular standard entropy of NO is given by:

$$S_{molar} = \frac{5}{2}R + R\ \ln(q_{electronic}) = \frac{5}{2}R + R\ \ln(g_0 + g_1\ \text{Exp}[-\beta\ \varepsilon_1])$$

$$= \frac{5}{2}R + R\ \ln\left(2 + 2\ \text{Exp}\left[-\frac{\left(6.626\times10^{-34}\ \text{J s}\right)\times\left(2.998\times10^{8}\ \text{m s}^{-1}\right)\times\left(12110\ \text{m}^{-1}\right)}{\left(1.3807\times10^{-23}\ \text{J K}^{-1}\right)\times(298\ \text{K})}\right]\right)$$

$$= \underline{\frac{5}{2}R + R\ \ln(3.11)}$$

Without the degeneracies the entropy would be:

$$S_{molar} = \frac{5}{2}R + R\ln\left(1 + 1\,\mathrm{Exp}\left[-\frac{(6.626\times10^{-34}\ \mathrm{J\,s})\times(2.998\times10^{8}\ \mathrm{m\,s^{-1}})\times(12110\,\mathrm{m^{-1}})}{(1.3807\times10^{-23}\ \mathrm{J\,K^{-1}})\times(298\,\mathrm{K})}\right]\right)$$

$$= \frac{5}{2}R + R\ln(1.56)$$

Therefore the degeneracies of the ground and first excited state of NO contribute:

$$R\ln(3.11) - R\ln(1.56) = R\,(\ln(3.11) - (1.56)) = R\left(\ln\left(\frac{(3.11)}{(1.56)}\right)\right) = R\ln(1.99) \cong R\ln(2)$$

A comparison with R ln(4) is valid since this would represent the contribution to the entropy if the probability of occupying all four states of NO would be equal. Due to the energy gap between the ground and excited state, the contribution is about half of that.

P23.42) Determine the equilibrium constant for the dissociation of sodium at 298 K: $Na_2(g) = \rightleftarrows 2Na(g)$. For Na_2, $B = 0.155\ \mathrm{cm^{-1}}$, $\tilde{\nu} = 159\ \mathrm{cm^{-1}}$, the dissociation energy is 70.4 kJ/mol, and the ground-state electronic degeneracy for Na is 2.

The equilibrium constant, K_p, for the reaction is given by:

$$K_p = \frac{\left(\frac{q(Na)}{N_A}\right)^2}{\left(\frac{q(Na_2)}{N_A}\right)}\mathrm{Exp}[\beta E_D]$$

We first calculate the partition functions. Since Na is a monoatomic species we only need to consider q_T and q_E, with $g_0 = 2$:

$$q_{total}(Na) = 2\times(2\pi\ m_{Na}\ kT)^{3/2}\frac{V}{h^3} = 2\times(2\pi\ m_{Na}\ kT)^{3/2}\frac{\left(\frac{RT}{pN_A}\right)}{h^3}$$

$$2\times\left(2\pi\,(3.81933\times10^{-26}\ \mathrm{kg})\times(1.3807\times10^{-23}\ \mathrm{J\,K^{-1}})\times(298\,\mathrm{K})\right)^{3/2}$$

$$\times\frac{\left(\frac{(8.314472\ \mathrm{J\,mol^{-1}\,K^{-1}})\times(298\,\mathrm{K})}{(101325\,\mathrm{Pa})\times(6.022\times10^{23}\ \mathrm{mol^{-1}})}\right)}{(6.626\times10^{-34}\ \mathrm{J\,s})^3} = 8.6615\times10^{6}$$

For Na_2 we also need to include the vibrational and rotational partition functions. With $\sigma = 2$, we obtain:

$$q_{total}(Na_2) = \left\{ \left(2\pi\ m_{Na_2}\ kT\right)^{3/2} \frac{\left(\frac{RT}{pN_A}\right)}{h^3} \right\} \times \left\{ \frac{kT}{\sigma hcB} \right\} \times \left\{ \frac{1}{\left(1 - Exp\left[-\frac{hc\tilde{\nu}}{kT}\right]\right)} \right\} \times \{1\}$$

$$= \left(2\pi\left(7.63866\times10^{-26}\ kg\right)\times\left(1.3807\times10^{-23}\ J\,K^{-1}\right)\times(298\ K)\right)^{3/2}$$

$$\times \frac{\left(\frac{\left(8.314472\ J\,mol^{-1}\ K^{-1}\right)\times(298\ K)}{(101325\ Pa)\times\left(6.022\times10^{23}\ mol^{-1}\right)}\right)}{\left(6.626\times10^{-34}\ J\,s\right)^3}$$

$$\times\left\{ \frac{\left(1.3807\times10^{-23}\ J\,K^{-1}\right)\times(2500\ K)}{2\times\left(6.626\times10^{-34}\ J\,s\right)\times\left(2.998\times10^8\ m\,s^{-1}\right)\times\left(15.5\ m^{-1}\right)} \right\}$$

$$\times\left\{ \frac{1}{\left(1 - Exp\left[-\frac{\left(6.626\times10^{-34}\ J\,s\right)\times\left(2.998\times10^8\ m\,s^{-1}\right)\times\left(15900\ m^{-1}\right)}{\left(1.3807\times10^{-23}\ J\,K^{-1}\right)\times(298\ K)}\right]\right)} \right\}$$

$$\times\{1\} = 1.5272\times10^{10}$$

Then:

$$K_p = \frac{\left(\frac{\left(8.6615\times10^6\right)}{\left(6.022\times10^{23}\right)}\right)^2}{\left(\frac{\left(1.5272\times10^{10}\right)}{\left(6.022\times10^{23}\right)}\right)} Exp\left[\frac{\left(70.4\ kJ\,mol^{-1}\right)}{\left(1.3807\times10^{-23}\ J\,K^{-1}\right)\times(298\ K)\times\left(6.022\times10^{23}\right)}\right] = \underline{1.78\times10^{-8}}$$

P23.46) The average length of a helical segment in a polypeptide is given by

$$\langle i \rangle = \sum_i i p_i$$

Using the expression for p_i from the zipper model, derive the expression for $\langle i \rangle$. For s = 1.6, σ = 10^{-3}, and N = 30, what is $\langle i \rangle$?

Using $p = 1 + \frac{\sigma s^2\left(s^N + Ns^{-1} - (N+1)\right)}{(s-1)^2}$

$$p = 1 + \frac{10^{-3}\times1.6^2\left(1.6^{30} + 30\times1.6^{-1} - (30+1)\right)}{(1.6-1)^2}$$

$$p = 9453.2$$

With $p_i = \dfrac{(N-i+1)\sigma s^i}{q}$:

$$\langle i \rangle = \sum_i i \frac{(N-i+1)\sigma s^i}{q} = \sum_{i=1}^{30} i \frac{(30-i+1)10^{-3} \times 1.6^i}{9453.2}$$

$$\underline{\langle i \rangle = 26.7}$$

Chapter 24: Transport Phenomena

P24.2)

a. The diffusion coefficient for Xe at 273 K and 1 atm is $0.5 \times 10^{-5}\ m^2\ s^{-1}$. What is the collisional cross section of Xe?

b. The diffusion coefficient of N_2 is threefold greater than that of Xe under the same pressure and temperature conditions. What is the collisional cross section of N_2?

a) The collisional cross section for Xe can be calculated as follows. Using:

$$D = \frac{1}{3}\left(\frac{8\,k\,T}{\pi\,m}\right)^{1/2}\left(\frac{R\,T}{\sqrt{2}\,p\,\sigma}\right),$$

and solving for r yields:

$$\sigma = \frac{kT\left(\dfrac{8\,k\,T}{\pi\,m_{Xe}}\right)^{1/2}}{3\sqrt{2}\,p\,D} = \frac{kT\left(\dfrac{8\times\left(1.3807\times10^{-23}\ J\ K^{-1}\right)\times\left(273\ K\right)}{\pi\ \left(\ 2.18017\times10^{-25}\ kg\right)}\right)^{1/2}}{3\sqrt{2}\ (101325\ Pa)\times\left(0.5\times10^{-5}\ m^2\ s^{-1}\right)} = \underline{\ 3.67955\times10^{-19}\ m^2}$$

a) The collisional cross section for N_2 with $D(N_2) = 3\times D(Xe)$ is:

$$\sigma = \frac{\left(\dfrac{8\,k\,T}{\pi\,m_{N_2}}\right)^{1/2}}{3\sqrt{2}\,p\,D} = \frac{\left(\dfrac{8\times\left(1.3807\times10^{-23}\ J\ K^{-1}\right)\times\left(273\ K\right)}{\pi\ \left(\ 4.65294\times10^{-26}\ kg\right)}\right)^{1/2}}{3\sqrt{2}\ (101325\ Pa)\times\left(1.5\times10^{-5}\ m^2\ s^{-1}\right)} = \underline{\ 2.65494\times10^{-19}\ m^2}$$

P24.6) Myoglobin is a protein that participates in oxygen transport. For myoglobin in water at 20 °C, $\bar{s} = 2.04 \times 10^{-13}$ s, $D = 11.3 \times 10^{-11}\ m^2\ s^{-1}$, and $\bar{V} = 0.740\ cm^3\ g^{-1}$. The density of water is 0.998 g cm^3 and the viscosity is 1.002 cP at this temperature.

a. Using the information provided, estimate the size of myoglobin.

b. What is the molecular weight of myoglobin?

a) The size of myoglobin can be estimated by calculating a molecular radius using the Stokes equation, and solving for r:

$$D = \frac{k\,T}{6\,\pi\,\mu\,r},$$

for r, resulting in:

$$r = \frac{k\,T}{6\,\pi\,\mu\,D} = \frac{\left(1.3807\times10^{-23}\ \text{J K}^{-1}\right)\times\left(298\ \text{K}\right)}{6\,\pi\,\left(1.002\times10^{-3}\ \text{kg s}^{-1}\ \text{m}^{-1}\right)\times\left(11.3\times10^{-11}\ \text{m}^{2}\ \text{s}^{-1}\right)} = \underline{1.896\times10^{-9}\ \text{m}}$$

b) To get the molecular weight of myoglobin we use:

$$m = \frac{\bar{s}\,f}{\left(1-\bar{V}\,\rho\right)},\ f = 6\pi\,\mu\,r,\ \text{and}\ M = \frac{m}{n},$$

to obtain:

$$M = \frac{\bar{s}\,6\pi\,\mu\,r\,N_A}{\left(1-\bar{V}\,\rho\right)n} = \frac{6\,\pi\left(1.002\times10^{-3}\ \text{kg s}^{-1}\ \text{m}^{-1}\right)\times\left(2.04\times10^{-13}\ \text{s}^{-1}\right)\times\left(1.896\times10^{-9}\ \text{m}\right)\times\left(6.022\times10^{23}\ \text{mol}^{-1}\right)}{\left(1-\left(0.740\,\text{cm}^3\ \text{g}^{-1}\right)\times\left(0.998\ \text{g cm}^{-3}\right)\right)}$$

$$= \underline{16.83\ \text{kg mol}^{-1}}$$

P24.8) Boundary centrifugation is performed at an angular velocity of 40,000 rpm to determine the sedimentation coefficient of cytochrome *c* (M = 13,400 g mol^{-1}) in water at 20 °C (ρ = 0.998 g cm^3, η = 1.002 cP). The following data are obtained on the position of the boundary layer as a function of time:

Time (h)	x_b **(cm)**
0	4.00
2.5	4.11
5.2	4.23
12.3	4.57
19.1	4.91

a. What is the sedimentation coefficient for cytochrome *c* under these conditions?

b. The specific volume of cytochrome *c* is 0.728 cm^3 g^{-1}. Estimate the size of cytochrome *c*.

a) We use the data to make a plot of $\ln\left(\frac{x_{b,t}}{x_{b,t=0}}\right)$ versus time:

time [s]	x_b [cm]	$\ln\left(\frac{x_{b,t}}{x_{b,t=0}}\right)$
0	4.00	0.0
9000	4.11	0.0271287
18720	4.23	0.0559076
44280	4.57	0.133219

68760	4.91	0.20498

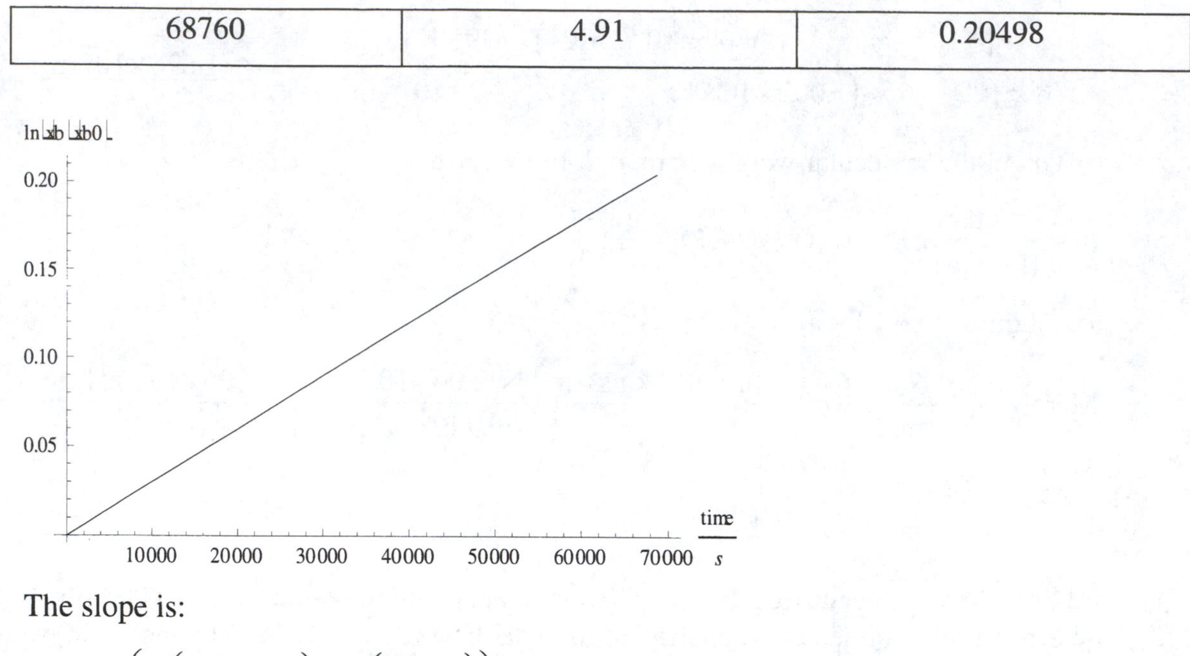

The slope is:

$$\text{slope} = \frac{\left(\ln\left(\frac{x_{b,t=68760s}}{x_{b,t=0s}}\right) - \ln\left(\frac{x_{b,t=0s}}{x_{b,t=0s}}\right)\right)}{\left(t_{68760} - t_0\right)} = \frac{(0.20498-0)}{(68760-0.0)} = 2.98109\times10^{-6}\ \text{s}^{-1}$$

With:

$$\omega^2\,\bar{s}\,t = \ln\left(\frac{x_{b,t}}{x_{b,t=0}}\right) = \text{slope}\ t\ ,$$

we obtain for the sedimentation coefficient:

$$\bar{s} = \frac{\text{slope}}{\omega^2} = \frac{\left(2.98109\times10^{-6}\ \text{s}^{-1}\right)}{\left[\left(40000\ \text{rev min}^{-1}\right)\times\left(2\pi\ \text{rad rev}^{-1}\right)\times\left(0.0167\ \text{min s}^{-1}\right)\right]^2} = \underline{1.69224\times10^{-13}\text{s}}$$

b) To estimate the size of cytochrome we calculate a radius for the molecule from:

$$m = \frac{\bar{s}\,6\pi\,\mu\,rN_A}{\left(1-\bar{V}\,\rho\right)} = MN_A,$$

and solve for r:

$$r = \frac{M\left(1-\bar{V}\,\rho\right)}{\bar{s}\,6\pi\,\mu\,N_A} = \left(\frac{(1\text{mol})\times\left(13400\times10^{-3}\text{kg mol}^{-1}\right)\times\left\{1-\left(0.728\,\text{cm}^3\ \text{g}^{-1}\right)\times\left(0.998\ \text{g cm}^{-3}\right)\right\}}{6\pi\times\left(1.002\times10^{-3}\text{kg s}^{-1}\ \text{m}^{-1}\right)\times\left(1.69\times10^{-13}\text{s}\right)\times\left(6.022\times10^{23}\ \text{mol}^{-1}\right)}\right)$$

$$\underline{=1.90\times10^{-9}\ \text{m}}$$

P24.10) In the early 1990s, fusion involving hydrogen dissolved in palladium at room temperature, or *cold fusion,* was proposed as a new source of energy. This process relies on the diffusion of H_2 into palladium. The diffusion of hydrogen gas through a 0.005-cm-thick piece of palladium foil with a cross section of 0.750 cm^2 is measured. On one side of the foil, a volume of gas maintained at 298 K and 1 atm is applied, while a vacuum is applied to the other side of the foil. After 24 h, the volume of hydrogen has decreased by 15.2 cm^3. What is the diffusion coefficient of hydrogen gas in palladium?

Using the expression for flux:

$$J_x = -D\left(\frac{d\tilde{N}(x)}{dx}\right)$$

If the flux and gradient in number density are determined, then the diffusion coefficient can be defined. The flux is equal to the number of particles that pass through the foil area per unit time. First the number of particles that diffused through the foil is determined using the ideal gas law:

$$n_{diffused} = \frac{(101325\ \text{Pa})\times(15.2\times10^{-6}\ \text{m}^3)}{(8.314472\text{J mol}^{-1}\ \text{K}^{-1})\times(298\ \text{K})} = 6.21\times10^{-4}\ \text{mol}$$

$$N_{diffused} = n_{diffused}N_A = 3.74\times10^{20}$$

Using the foil area and the time for diffusion, the flux is determined as follows:

$$J_x = \frac{N_{diffused}}{A\,t} = \frac{(3.74\times10^{20})}{(0.750\times10^{-4}\ \text{m}^2)\times(24\ \text{h})\times\left(\frac{3600\ \text{s}}{1\ \text{h}}\right)} = 5.77\times10^{19}\ \text{m}^{-2}\ \text{s}^{-1}$$

Next the gradient in number density is equal to the difference in gas number density between each side of the foil divided by the foil thickness. The number density in the incident side of the foil is determined using the ideal gas law:

$$\tilde{N} = \frac{P}{kT} = \frac{(101325\ \text{Pa})}{(1.38\times10^{-23}\ \text{J K}^{-1})\times(298\ \text{K})} = 2.46\times10^{25}\ \text{m}^{-3},$$

and the spatial gradient in number density is:

$$\frac{d\tilde{N}}{dx} = \frac{(0 - 2.46\times10^{25})\text{m}^{-3}}{(5\times10^{-5}\ \text{m})} = -4.93\times10^{29}\ \text{m}^{-4}$$

Finally, the diffusion coefficient is calculated as:

$$D = -\frac{J_x}{\left(\frac{d\tilde{N}}{dx}\right)} = \frac{\left(5.77\times10^{19}\ \text{m}^{-2}\ \text{s}^{-1}\right)}{\left(-4.93\times10^{29}\ \text{m}^{-4}\right)} = \underline{1.17\times10^{-10}\ \text{m}^2\ \text{s}^{-1}}$$

P24.12)

a. The viscosity of O_2 at 293 K and 1 atm is 204 μP. What is the expected flow rate through a tube having a radius of 2 mm, length of 10 cm, input pressure of 765 torr, output pressure of 760 torr, with the flow measured at the output end of the tube?

b. If Ar were used in the apparatus ($\eta = 223\ \mu$P) of part (a), what would be the expected flow rate? Can you determine the flow rate without evaluating Poiseuille's equation?

We use Poiseuille's law:

$$\frac{\Delta V}{\Delta t} = \frac{\pi r^4}{8\eta}\left(\frac{(p_{in} - p_{out})}{(\ell)}\right)$$

a)

$$\frac{\Delta V}{\Delta t} = \frac{\pi\left(2\times10^{-3}\text{m}\right)^4}{8\left(204\times10^{-6}\text{kg m}^{-1}\text{s}^{-1}\right)}\left(\frac{\left(765\ \text{Torr}\frac{133.32\ \text{Pa}}{1\ \text{Torr}} - 760\ \text{Torr}\frac{133.32\ \text{Pa}}{1\ \text{Torr}}\right)}{(0.1\ \text{m})}\right) = \underline{2.05\times10^{-4}\ \text{m}^3\ \text{s}^{-1}}$$

b)

$$\frac{\Delta V}{\Delta t} = \frac{\pi\left(2\times10^{-3}\text{m}\right)^4}{8\left(223\times10^{-6}\text{kg m}^{-1}\text{s}^{-1}\right)}\left(\frac{\left(765\ \text{Torr}\frac{133.32\ \text{Pa}}{1\ \text{Torr}} - 760\ \text{Torr}\frac{133.32\ \text{Pa}}{1\ \text{Torr}}\right)}{(0.1\ \text{m})}\right) = \underline{1.88\times10^{-4}\ \text{m}^3\ \text{s}^{-1}}$$

O_2 and Ar could be treated as ideal gases:

P24.14) The viscosity of H_2 at 273 K at 1 atm is 84 μP. Determine the viscosities of D_2 and HD.

The viscosity of a gas is given by:

$$\eta = \frac{1}{3}\left(\frac{8RT}{\pi M}\right)^{1/2}\tilde{N}\left(\frac{RT}{\sqrt{2}\,p\sigma}\right)m$$

From this we deduce that for the viscosity of a gas:

$\eta \propto M$

Therefore:

$$\frac{\eta_{D_2}}{\eta_{H_2}} = \frac{M_{D_2}^{1/2}}{M_{H_2}^{1/2}}$$

$$\eta_{D_2} = \eta_{H_2}\frac{M_{D_2}^{1/2}}{M_{H_2}^{1/2}} = \left(84\times10^{-6}\,\text{P}\right)\sqrt{\frac{\left(4.0\,\text{g mol}^{-1}\right)}{\left(2.0\,\text{g mol}^{-1}\right)}} = \underline{118.8\,\mu\text{P}}$$

$$\eta_{HD} = \eta_{H_2}\frac{M_{D_2}^{1/2}}{M_{H_2}^{1/2}} = \left(84\times10^{-6}\,\text{P}\right)\sqrt{\frac{\left(3.0\,\text{g mol}^{-1}\right)}{\left(2.0\,\text{g mol}^{-1}\right)}} = \underline{102.9\,\mu\text{P}}$$

P24.21) A more exact formulation of the frictional coefficient of a cylinder is $F_{tr} = \frac{f_{tr}}{f_{tr}^{\circ}} \approx \frac{\left(2/3\right)^{1/3} P^{2/3}}{\ln\left(P\right)+\gamma}$ where $\gamma = 0.312 + \frac{0.565}{P} + \frac{0.100}{P^2}$ is intended to account for the abrupt ends of the cylinder. A hydrated DNA oligomer can be treated as a rodlike molecule of length 20.00 nm and diameter 2.00 nm. Calculate f_{tr}, f_{tr}°, and the Perrin factor. Assume $\eta = 0.891$ cP. Also calculate the translational diffusion coefficient at $T =$ 298 K. Compare this result to the result obtained in Example Problem 24.8.

We first have to calculate the radius, re, for which the volume of the sphere is equal to the volume of the rod:

$$V_{sphere} = \frac{4\,\pi}{3} r_e^{\,3} \text{ and } V_{rod} = \pi\left(\frac{d}{3}\right)^2 L$$

Then:

$$r_e = \left(\frac{3\,d^2\,L}{16}\right)^{1/3} = \left(\frac{3\times\left(2.0\times10^{-9}\,\text{m}\right)^2\times\left(20.0\times10^{-9}\,\text{m}\right)}{16}\right)^{1/3} = 2.46621\times10^{-9}\,\text{m}$$

Now we can calculate f_{tr}^{0}, f_{tr}, and the Perrin Factor, F_{tr}, with P as the ratio of the major and minor axis (20 nm / 2 nm = 10 nm):

$$f_{tr}^{\,0} = 6\,\pi\,\eta\,r_e = 6\,\pi\left(0.891\times10^{-3}\,\text{kg s}^{-1}\,\text{m}^{-1}\right)\times\left(2.46621\times10^{-9}\,\text{m}\right) = \underline{4.14199\times10^{-11}\,\text{kg s}^{-1}}$$

$$f_{tr} = f_{tr}^{\,0}\frac{P^{-1/3}\sqrt{P^2-1}}{\ln\left(P+\sqrt{P^2-1}\right)} = 4.14199\times10^{-11}\,\text{kg s}^{-1}\,\frac{10^{-1/3}\sqrt{10^2-1}}{\ln\left(10+\sqrt{10^2-1}\right)} = \underline{6.28534\times10^{-11}\,\text{kg s}^{-1}}$$

$$F_{tr} = \frac{f_{tr}}{f_{tr}^{0}} = \frac{\left(\frac{2}{3}\right)^{-1/3} P^{2/3}}{\ln(P) + \left(0.312 + \frac{0.565}{P} + \frac{0.100}{P^2}\right)} = \frac{\left(\frac{2}{3}\right)^{-1/3} 10^{2/3}}{\ln(10) + \left(0.312 + \frac{0.565}{10} + \frac{0.100}{10^2}\right)} = \underline{1.52}$$

And finally the diffusion coefficient:

$$D_{tr} = \frac{k_B T}{f_{tr}} = \frac{\left(1.3807 \times 10^{-23}\ \text{J K}^{-1}\right) \times (298\ \text{K})}{\left(6.28534 \times 10^{-11}\ \text{kg s}^{-1}\right)} = \underline{6.54617 \times 10^{-11}\ \text{m}^2\ \text{s}^{-1}}$$

The diffusion coefficient calculated in the example problem differs from the more exact coefficient calculated in this problem by:

$$\text{diviation} = \left(1 - \frac{6.40 \times 10^{-11}\ \text{m}^2\ \text{s}^{-1}}{6.54617 \times 10^{-11}\ \text{m}^2\ \text{s}^{-1}}\right) \times 100 = \underline{2.23\%}$$

Chapter 25: Elementary Chemical Kinetics

P25.2) Consider the first-order decomposition of cyclobutane at 438 °C at constant volume:

$$C_4H_8(g) \longrightarrow 2C_2H_4(g)$$

a. Express the rate of the reaction in terms of the change in total pressure as a function of time.

b. The rate constant for the reaction is $2.48 \times 10^{-4}\ s^{-1}$. What is the half-life?

c. After initiation of the reaction, how long will it take for the initial pressure of C_4H_8 to drop to 90% of its initial value?

a) The rate for the reaction expressed as a function of the total pressure is:

$$\text{rate} = -\frac{dn_{C_4H_8}}{dt} = -p_{tot}\frac{dp_{C_4H_8}}{dt}$$

b) For a first-order reaction, the half life is:

$$t_{1/2} = \frac{\ln(2)}{k_{rate}} = \frac{\ln(2)}{(2.48\times10^{-4}\ s^{-1})} = \underline{2.79\times10^{3}\ s}$$

c) To calculate the time we use the equation for a first-order rate:

$$[A] = [A]_0 e^{-k_{rate}t}$$

We set $[A] = 0.9[A]_0$ and solve for t:

$$t_{0.9} = -\frac{\ln(0.9)}{k_{rate}} = \frac{\ln(0.9)}{(2.48\times10^{-4}\ s^{-1})} = \underline{424.8\ s}$$

P25.7) The reaction rate as a function of initial reactant pressures was investigated for the reaction $2NO(g) + 2H_2(g) \longrightarrow N_2(g) + 2H_2O(g)$, and the following data were obtained:

Run	P_o **H_2 (kPa)**	P_o **NO (kPa)**	**Rate (kPa s^{-1})**
1	53.3	40.0	0.137
2	53.3	20.3	0.033
3	38.5	53.3	0.213
4	19.6	53.3	0.105

What is the rate law expression for this reaction?

The order of the reaction with respect to a given reactant, α, is given by:

$$\ln\left(\frac{\boldsymbol{R}_1}{\boldsymbol{R}_2}\right) = \alpha \ln\left(\frac{[A]_1}{[A]_2}\right)$$

$$\alpha = \frac{\ln\left(\frac{\boldsymbol{R}_1}{\boldsymbol{R}_2}\right)}{\ln\left(\frac{[A]_1}{[A]_2}\right)}$$

We use rows 1 and 2 in the table to obtain the order with respect to NO, since the partial pressure of H_2 was held constant:

$$\alpha_{NO} = \frac{\ln\left(\frac{(0.137\,\text{kPa})}{(0.033\,\text{kPa})}\right)}{\ln\left(\frac{(40.0\,\text{kPa s}^{-1})}{(20.3\,\text{kPa s}^{-1})}\right)} = 2.10 \approx 2$$

We use rows 3 and 4 in the table to obtain the order with respect to H_2, since the partial pressure of NO was held constant:

$$\alpha_{H_2} = \frac{\ln\left(\frac{(0.213\,\text{kPa})}{(0.103\,\text{kPa})}\right)}{\ln\left(\frac{(38.05\,\text{kPa s}^{-1})}{(19.6\,\text{kPa s}^{-1})}\right)} = 1.05 \approx 1$$

The rate law expression is:

$$R = k_{rate}\,[NO]^2\,[H_2]$$

We obtain the rate constant from any of the data in the four rows in the table:

$$k_{rate} = \frac{R}{[NO]^2\,[H_2]} = \frac{(0.137\,\text{kPa s}^{-1})}{[40.0\,\text{kPa}]^2\,[53.3\,\text{kPa}]} = 1.6\times10^{6}\ \text{kPa s}^{-1}$$

P25.12) The half-life of ^{238}U is 4.5×10^9 years. How many disintegrations occur in 1 min for a 10-mg sample of this element?

To calculate the number of disintegrations we use the first-order rate equation:

$$[A] = [A]_0 e^{-k_{rate}\,t}$$

$$\ln[A] = \ln[A]_0 - k_{rate}\,t$$

First we calculate the number of ^{238}U nuclei for a 10-mg sample:

$$\text{nuclei} = \frac{m}{M}N_A = \frac{(10\times10^{-3}\ \text{g})}{(238\ \text{g mol}^{-1})}\times(6.022\times10^{23}\ \text{mol}^{-1}) = 2.53\times10^{19}$$

Now we need to calculate the rate constant:

$$k_{rate} = \frac{\ln(2)}{t_{1/2}} = \frac{\ln(2)}{(1.43\times10^{17}\ \text{s})} = \underline{4.88\times10^{-18}\ \text{s}^{-1}}$$

The number of nuclei still present after 1 min is:

$$\ln[A] = \ln(2.53\times10^{19}) - (4.88\times10^{-18}\ \text{s}^{-1})(60\ \text{s}) = 44.677$$

$$[A] = \underline{2.53\times10^{19}}$$

1 min is too short of a decay time to produce a detectible number of disintegrations.

P25.16) Show that the ratio of the half-life to the three-quarter life, $t_{1/2}/t_{3/4}$, for a reaction that is *n*th order ($n > 1$) in reactant A can be written as a function of *n* alone (that is, there is no concentration dependence in the ratio).

In general for order n:

$$k\,t = \frac{1}{(n-1)}\left\{\frac{1}{[A]^{n-1}} - \frac{1}{[A]_0^{n-1}}\right\}$$

With $[A] = \frac{1}{2}[A]_0$ for $t_{1/2}$ we obtain:

$$k\,t_{1/2} = \frac{1}{(n-1)}\left\{\frac{2^{n-1}}{[A]_0^{n-1}} - \frac{1}{[A]_0^{n-1}}\right\}$$

$$t_{1/2} = \frac{(2^{n-1}-1)}{k\,(n-1)[A]_0^{n-1}}$$

With $[A] = \frac{1}{4}[A]_0$ for $t_{3/4}$ we obtain:

$$k\,t_{3/4} = \frac{1}{(n-1)}\left\{\frac{4^{n-1}}{[A]_0^{n-1}} - \frac{1}{[A]_0^{n-1}}\right\}$$

$$t_{3/4} = \frac{(4^{n-1}-1)}{k\,(n-1)[A]_0^{n-1}}$$

And:

$$\frac{t_{1/2}}{t_{3/4}} = \frac{\left(\dfrac{(2^{n-1}-1)}{k(n-1)[A]_0^{n-1}}\right)}{\left(\dfrac{(4^{n-1}-1)}{k(n-1)[A]_0^{n-1}}\right)} = \frac{(2^{n-1}-1)}{(4^{n-1}-1)}$$

P25.18) For the sequential reaction $A \xrightarrow{k_A} B \xrightarrow{k_B} C$, , the rate constants are $k_A = 5 \times 10^6\ s^{-1}$ and $k_B = 3 \times 10^6\ s^{-1}$. Determine the time at which [B] is at a maximum.

The time for which [B] is at the maximum is:

$$t_{max} = \frac{1}{t_A - t_B}\ln\left(\frac{t_A}{t_B}\right) = \frac{1}{\left((5\times10^6\ s^{-1})-(3\times10^6\ s^{-1})\right)}\ln\left(\frac{(5\times10^6\ s^{-1})}{(3\times10^6\ s^{-1})}\right) = \underline{2.55\times10^{-7}s}$$

P25.21) For a type II second-order reaction, the reaction is 60% complete in 60 seconds when $[A]_0 = 0.1$ M and $[B]_0 = 0.5$ M.

a. What is the rate constant for this reaction?

b. Will the time for the reaction to reach 60% completion change if the initial reactant concentrations are decreased by a factor of 2?

We use:

$$\frac{1}{[B]_0-[A]_0}\ln\left(\frac{[B]/[B]_0}{[A]/[A]_0}\right) = k\,t$$

a) The concentrations for the reactants after 60 s when the reaction is 60% complete are:

$$[B] = [B]_0-(1\text{-}0.6)[A]_0 = (0.5\,M)\text{-}(1\text{-}0.6)\times(0.1\,M) = 0.46\,M$$
$$[A] = (1\text{-}0.6)[A]_0 = (1\text{-}0.6)\times(0.1\,M) = 0.04\,M$$

The rate constant is then:

$$k = \frac{\dfrac{1}{[B]_0-[A]_0}\ln\left(\dfrac{[B]/[B]_0}{[A]/[A]_0}\right)}{t} = \frac{\dfrac{1}{(0.5\,M)-(0.1\,M)}\ln\left(\dfrac{(0.46\,M)/(0.5\,M)}{(0.04\,M)/(0.1\,M)}\right)}{(60\,s)} = \underline{0.0347\,s^{-1}\,M^{-1}}$$

b) The concentrations for the reactants after 60 s when the reaction is 60% complete are:

$$[B] = [B]_0-(1\text{-}0.6)[A]_0 = (1.0\,M)\text{-}(1\text{-}0.6)\times(0.2\,M) = 0.992\,M$$
$$[A] = (1\text{-}0.6)[A]_0 = (1\text{-}0.6)\times(0.2\,M) = 0.008\,M$$

The rate constant is then:

$$t = \frac{\frac{1}{[B]_0-[A]_0}\ln\left(\frac{[B]/[B]_0}{[A]/[A]_0}\right)}{k} = \frac{\frac{1}{(1.0\,M)-(0.2\,M)}\ln\left(\frac{(0.992\,M)/(1.0\,M)}{(0.008\,M)/(0.2\,M)}\right)}{(0.0347\,s^{-1}\,M^{-1})} = \underline{115.7\,s}$$

P25.24) In the stratosphere, the rate constant for the conversion of ozone to molecular oxygen by atomic chlorine is $Cl + O_3 \longrightarrow ClO + O_2$ [$k = (1.7 \times 10^{10}\ M^{-1}\ s^{-1})e^{-260\ K/T}$].

a. What is the rate of this reaction at 20 km where [Cl] = 5×10^{-17} M, [O_3] = 8×10^{-9} M, and T = 220 K?

b. The actual concentrations at 45 km are [Cl] = 3×10^{-15} M and [O_3] = 8×10^{-11} M. What is the rate of the reaction at this altitude where T = 270 K?

c. (Optional) Given the concentrations in part (a), what would you expect the concentrations at 45 km to be assuming that the gravity represents the operative force defining the potential energy?

a) The rate of the reaction can be calculated from:

$$\text{rate} = k_{rate}[Cl][O_3]$$

$$= (5\times10^{-17}\ M)\times(8\times10^{-9}\ M)\times(1.7\times10^{10}\ M^{-1}\ s^{-1})\times \text{Exp}\left[-\frac{(260\ K)}{(220\ K)}\right] = \underline{2.09\times10^{-15}\ \ M\,s^{-1}}$$

b) For b) we obtain in analogy:

$$\text{rate} = k_{rate}[Cl][O_3]$$

$$= (3\times10^{-15}\ M)\times(8\times10^{-11}\ M)\times(1.7\times10^{10}\ M^{-1}\ s^{-1})\times \text{Exp}\left[-\frac{(260\ K)}{(270\ K)}\right] = \underline{1.56\times10^{-15}\ \ M\,s^{-1}}$$

P25.26) A standard "rule of thumb" for thermally activated reactions is that the reaction rate doubles for every 10 K increase in temperature. Is this statement true independent of the activation energy (assuming that the activation energy is positive and independent of temperature)?

The statement is not true; the ratio of rate constants depends on E_a:

$$\frac{k_1}{k_2} = \frac{A}{A}\frac{\text{Exp}\left[-\frac{E_a}{R\,T_1}\right]}{\text{Exp}\left[-\frac{E_a}{R\,T_2}\right]} = \text{Exp}\left[-\frac{E_a}{R\,T_1}\right]\text{Exp}\left[\frac{E_a}{R\,T_2}\right] = \underline{\text{Exp}\left[-\frac{E_a}{R}\left(\frac{1}{T_2}-\frac{1}{T_1}\right)\right]}$$

P25.30) Consider the reaction

$$A + B \underset{k'}{\overset{k}{\rightleftarrows}} P$$

A temperature-jump experiment is performed where the relaxation time constant is measured to be 310 μs, resulting in an equilibrium where $K_{eq} = 0.7$ with $[P]_{eq} = 0.2$ M. What are k and k'? (Watch the units!)

The constants k and k' can be calculated by using the two following equations for the relaxation time and overall equilibrium constant, respectively:

$$\tau = \frac{1}{k + k'} \text{ and } K_{eq} = \frac{k}{k'}$$

Solving both equations for k and setting them equal to each other yields:

$$\frac{1}{\tau} - k' = K_{eq}\, k'$$

Now solving for k' results in:

$$k' = \frac{1}{\tau(K_{eq} + 1)} = \frac{1}{(310 \times 10^{-6}\ \text{s})(0.7 + 1)} = \underline{1895.5\ \text{s}^{-1}}$$

We get k from:

$$k = K_{eq}\, k' = (0.7) \times (1895.5\ \text{s}^{-1}) = \underline{1328.3\ \text{s}^{-1}}$$

P25.32) In the following chapter, enzyme catalysis reactions will be extensively reviewed. The first step in these reactions involves the binding of a reactant molecule (referred to as a substrate) to a binding site on the enzyme. If this binding is extremely efficient (that is, equilibrium strongly favors the enzyme–substrate complex over separate enzyme and substrate) and the formation of product rapid, then the rate of catalysis could be diffusion limited. Estimate the expected rate constant for a diffusion-controlled reaction using typical values for an enzyme ($D = 1 \times 10^{-7}$ cm^2 s^{-1} and $r = 40$ Å) and a small molecular substrate ($D = 1 \times 10^{-5}$ cm^2 s^{-1} and $r = 5$ Å).

The rate constant for the complex formation in the diffusion limit can be estimated with:

$$\begin{aligned} k_d &= 4\pi N_A (r_A + r_B)(D_A + D_B) \\ &= 4\pi(6.022 \times 10^{23}\ \text{mol}^{-1}) \times ((5 \times 10^{-8}\ \text{cm}) + (40 \times 10^{-8}\ \text{cm})) \times ((1 \times 10^{-7}\ \text{cm}^2\text{s}^{-1}) + (1 \times 10^{-5}\ \text{cm}^2\text{s}^{-1})) \\ &= \underline{3.44 \times 10^{13}\ \text{cm}^3\ \text{s}^{-1}\ \text{mol}^{-1}} \end{aligned}$$

P25.35) The unimolecular decomposition of urea in aqueous solution is measured at two different temperatures and the following data are observed:

Trial Number	**Temperature (°C)**	k **(s^{-1})**
1	60.0	1.2×10^{-7}
2	71.5	4.40×10^{-7}

a. Determine the Arrhenius parameters for this reaction.

b. Using these parameters, determine $\Delta H^{\ddagger}$ and $\Delta S^{\ddagger}$ as described by the Eyring equation.

a) To determine the Arrhenius parameters from the data we plot ln(k) versus 1/T. The slope is then $-E_a/R$:

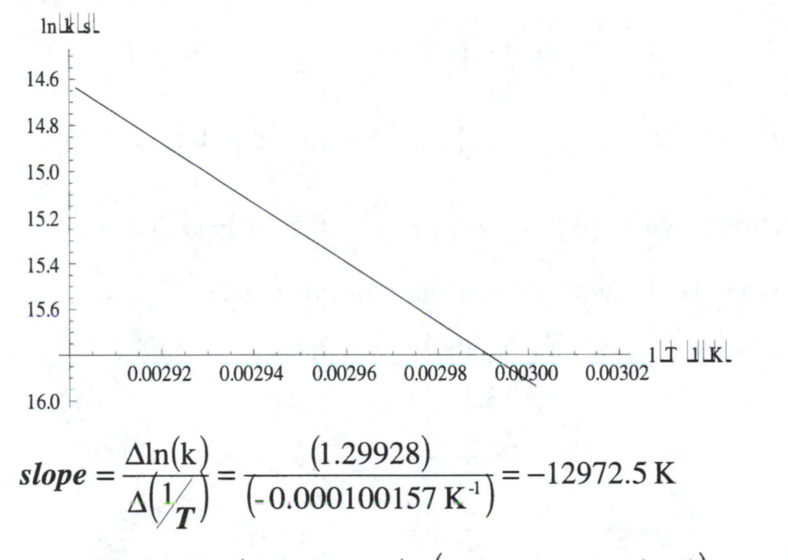

$$slope = \frac{\Delta \ln(k)}{\Delta\left(\frac{1}{T}\right)} = \frac{(1.29928)}{(-0.000100157\ K^{-1})} = -12972.5\ K$$

$$E_a = -slope\ R = (-12972.5\ K)\times(8.314472\ J\ mol^{-1}\ K^{-1}) = \underline{107.9\ kJ\ mol^{-1}}$$

The Arrhenius parameter, A, is then:

$$A = Exp\left[\ln(k) + \frac{E_a}{R\,T}\right] = Exp\left[\ln(1.2\times10^{-7}) + \frac{(107.9\ kJ\ mol^{-1})}{(333.15\ K)\times(8.314472\ J\ mol^{-1}\ K^{-1})}\right] = \underline{9.78\times10^{9}}$$

b) To obtain $\Delta H^{\#}$ we use the following equation:

$$E_a = \Delta H^{\#} + R\,T$$

$$\Delta H^{\#} = E_a - R\,T = (107.9\ kJ\ mol^{-1}) - (333.15\ K)\times(8.314472\ J\ mol^{-1}\ K^{-1}) = \underline{105.1\ kJ\ mol^{-1}}$$

And then finally $\Delta S^{\#}$ is given by:

$$A = \frac{e\,k_b\,T}{h\,c^0}\,\text{Exp}\left[\frac{\Delta S^{\#}}{R}\right]$$

$$\Delta S^{\#} = R\,\ln\left(\frac{A\,h\,c^0}{e\,k_b T}\right)$$

$$= \left(8.314472\,\text{J mol}^{-1}\,\text{K}^{-1}\right)\times\ln\left(\frac{\left(9.78\times10^{9}\right)\times\left(6.626\times10^{-34}\,\text{J s}\right)\times\left(1\,\text{M}\right)}{e\times\left(1.3807\times10^{-23}\,\text{J K}^{-1}\right)\times\left(333.15\,\text{K}\right)}\right) = \underline{-62.9\,\text{J mol}^{-1}\,\text{K}^{-1}}$$

P25.39) Chlorine monoxide (ClO) demonstrates three bimolecular self-reactions:

$$Rxn_1:\ \text{ClO}\cdot(g) + \text{ClO}\cdot(g) \xrightarrow{k_1} \text{Cl}_2(g) + \text{O}_2(g)$$

$$Rxn_2:\ \text{ClO}\cdot(g) + \text{ClO}\cdot(g) \xrightarrow{k_2} \text{Cl}\cdot(g) + \text{ClOO}\cdot(g)$$

$$Rxn_3:\ \text{ClO}\cdot(g) + \text{ClO}\cdot(g) \xrightarrow{k_3} \text{Cl}\cdot(g) + \text{OClO}\cdot(g)$$

The following table provides the Arrhenius parameters for this reaction:

	A **($M^{-1}\,s^{-1}$)**	***E_a*** **(kJ/mol)**
Rxn_1	6.08×10^8	13.2
Rxn_2	1.79×10^{10}	20.4
Rxn_3	2.11×10^8	11.4

a. For which reaction is $\Delta H^{\ddagger}$ greatest and by how much relative to the next closest reaction?

b. For which reaction is $\Delta S^{\ddagger}$ the smallest and by how much relative to the next closest reaction?

a) For a bimolecular gas reaction:

$E_a = \Delta H^{\#} + 2\,R\,T$ and $\Delta H^{\#} = E_a - 2\,R\,T$

Therefore, the larger the activation energy the larger $\Delta H^{\#}$ for a given reaction:

$\Delta H^{\#}(\text{Rxn2}) > \Delta H^{\#}(\text{Rxn1}) > \Delta H^{\#}(\text{Rxn3})$

$\Delta H^{\#}(\text{Rxn2})$ is 7200 kJ mol^{-1} larger than $\Delta H^{\#}(\text{Rxn1})$

b) For a bimolecular gas reaction:

$$A = \frac{e^2\, k_b\, T}{h\, c^0} \text{Exp}\left[\frac{\Delta S^{\#}}{R}\right]$$

$$\Delta S^{\#} = R \ln\left(\frac{A\, h\, c^0}{e^2\, k_b T}\right)$$

Therefore, the larger the Arrhenius parameter, A, the larger $\Delta S^{\#}$ for a given reaction:

$\Delta S^{\#}$ (Rxn3) < $\Delta S^{\#}$ (Rxn1) < $\Delta S^{\#}$ (Rxn2)

Setting:

$$\Delta S^{\#} = R \ln(A \text{ const})$$

We obtain the difference between the smallest and second smallest $\Delta S^{\#}$:

$$\Delta S^{\#}(3) - \Delta S^{\#}(1) = R \ln(A_3 \text{ const}) - R \ln(A_1 \text{ const}) = R \left(\ln(A_3 \text{ const}) - \ln(A_1 \text{ const})\right)$$

$$= R \left(\left(\ln(A_3) + \ln(\text{const})\right) - \left(\ln(A_1) + \ln(\text{const})\right)\right) = R \ln\left(\frac{A_3}{A_1}\right) = R\left(\ln\left(\frac{(2.11\times 10^8)}{(6.08\times 10^8)}\right)\right) = \underline{-1.06\, R}$$

Chapter 26: Complex Biological Reactions

P26.1) A proposed mechanism for the formation of N_2O_5 from NO_2 and O_3 is

$$NO_2 + O_3 \xrightarrow{k_1} NO_3 + O_2$$

$$NO_2 + NO_3 + M \xrightarrow{k_2} N_2O_5 + M$$

Determine the rate law expression for the production of N_2O_5 given this mechanism.

The rate overall reaction is represented by:

$$2\,NO_2 + O_3 \longrightarrow N_2O_5 + O_2$$

For the second mechanistic reaction equation the rate expression is:

$$R = \frac{d[N_2O_5]}{dt} = k_2\,[NO_3][NO_2][M]$$

Applying the steady-state approximation for the intermediate $[NO_3]$ yields:

$$\frac{d[NO_3]}{dt} = 0 = k_1\,[NO_2][O_3] - k_2\,[NO_2][NO_3][M]$$

Solving for $[NO_3]$:

$$k_1\,[NO_2][O_3] = k_2\,[NO_2][NO_3][M]$$

$$[NO_3] = \frac{k_1}{k_2}\frac{[O_3]}{[M]}$$

Substituting this result into the rate expression gives:

$$R = \frac{d[N_2O_5]}{dt} = k_2\,[NO_3][NO_2][M] = k_2\,\frac{k_1}{k_2}\frac{[O_3]}{[M]}[NO_2][M] = k_1\,[O_3][NO_2]$$

P26.4) The hydrogen–bromine reaction corresponds to the production of HBr from H_2 and Br_2 as follows: $H_2 + Br_2 \rightleftarrows 2HBr$. This reaction is famous for its complex rate law, determined by Bodenstein and Lind in 1906:

$$\frac{d[HBr]}{dt} = \frac{k[H_2][Br_2]^{1/2}}{1 + \dfrac{m[HBr]}{[Br_2]}}$$

where k and m are constants. It took 13 years for the correct mechanism of this reaction to be proposed, and this feat was accomplished simultaneously by Christiansen, Herzfeld, and Polyani. The mechanism is as follows:

$$Br_2 \underset{k_{-1}}{\overset{k_1}{\rightleftharpoons}} 2Br\cdot$$

$$Br\cdot + H_2 \overset{k_2}{\rightleftharpoons} HBr + H\cdot$$

$$H\cdot + Br_2 \overset{k_3}{\rightleftharpoons} HBr + Br\cdot$$

$$HBr + H\cdot \overset{k_4}{\rightleftharpoons} H_2 + Br\cdot$$

Construct the rate law expression for the hydrogen–bromine reaction by performing the following steps:

a. Write down the differential rate expression for [HBr].

b. Write down the differential rate expressions for [Br] and [H].

c. Because Br and H are reaction intermediates, apply the steady-state approximation to the result of part (b).

d. Add the two equations from part (c) to determine [Br] in terms of [Br_2].

e. Substitute the expression for [Br] back into the equation for [H] derived in part (c) and solve for [H].

f. Substitute the expressions for [Br] and [H] determined in part (e) into the differential rate expression for [HBr] to derive the rate law expression for the reaction.

a) $\frac{d[HBr]}{dt} = -k_4 [HBr][H] + k_2 [Br][H_2] + k_3 [Br_2][H]$

b) $\frac{d[H]}{dt} = k_2 [Br][H_2] - k_3 [Br_2][H] - k_4 [HBr][H]$

$\frac{d[Br]}{dt} = K[Br_2]^{1/2} - k_2 [Br][H_2] + k_3 [Br_2][H] + k_4 [HBr][H]$

c) $k_2 [Br][H_2] - k_3 [Br_2][H] - k_4 [HBr][H] = 0$

$-2k_{-1}[Br_2]^2 - k_2 [Br][H_2] + k_3 [Br_2][H] + k_4 [HBr][H] + 2k_1[Br] = 0$

d) Adding the steady-state equations yields:

$2k_1[Br] - 2k_{-1}[Br_2]^2 = 0$

$[Br] = \left(\frac{k_1}{k_{-1}}[Br_2]\right)^{1/2}$

e) Substituting into the equation for [H] gives:

$$[H]=\frac{k_2\ [H_2][Br]}{(k_3\ [Br_2]+k_4\ [HBr])}=\frac{k_2\ [H_2]}{(k_3\ [Br_2]+k_4\ [HBr])}\left(\frac{k_1}{k_{-1}}[Br_2]\right)^{1/2}=\frac{k_2\left(\frac{k_1}{k_{-1}}\right)^{1/2}[Br_2]^{1/2}[H_2]}{(k_3\ [Br_2]+k_4\ [HBr])}$$

f) $\frac{d[HBr]}{dt}=-k_4\ [HBr][H]+k_2\ [Br][H_2]+k_3\ [Br_2][H]$

$$=-k_4\ [HBr]\frac{k_2\left(\frac{k_1}{k_{-1}}\right)^{1/2}[Br_2]^{1/2}[H_2]}{(k_3\ [Br_2]+k_4\ [HBr])}+k_2\left(\frac{k_1}{k_{-1}}[Br_2]\right)^{1/2}[H_2]+k_3\ [Br_2]\frac{k_2\left(\frac{k_1}{k_{-1}}\right)^{1/2}[Br_2]^{1/2}[H_2]}{(k_3\ [Br_2]+k_4\ [HBr])}$$

Collecting terms and simplifying yields:

$$\frac{d[HBr]}{dt}=\frac{k'[Br_2]^{1/2}[H_2]}{\left(1+\frac{m[HBr]}{[Br_2]}\right)},\ \text{with } k'=k_2\left(\frac{k_1}{k_{-1}}\right)^{1/2},\ \text{and } m=\frac{k_4}{k_3}$$

P26.6) For the reaction I^- (*aq*) + OCl^- (*aq*)$^+$ OI^- (*aq*) + Cl^-(*aq*) occurring in aqueous solution, the following mechanism has been proposed:

$$OCl^- + H_2O \underset{k_{-1}}{\overset{k_1}{\rightleftharpoons}} HOCl + OH^-$$

$$I^- + HOCl \xrightarrow{k_2} HOI + Cl^-$$

$$HOI + OH^- \xrightarrow{k_3} H_2O + OI^-$$

a. Derive the rate law expression for this reaction based on this mechanism. (*Hint:* $[OH^-]$ should appear in the rate law.)

b. The initial rate of reaction was studied as a function of concentration by Chia and Connick [*J. Physical Chemistry* 63 (1959), 1518], and the following data were obtained:

$[I^-]_0$ (M)	$[OCl^-]_0$ (M)	$[OH^-]_0$ (M)	Initial Rate (M s^{-1})
2.0×10^{-3}	1.5×10^{-3}	1.00	1.8×10^{-4}
4.0×10^{-3}	1.5×10^{-3}	1.00	3.6×10^{-4}
2.0×10^{-3}	3.0×10^{-3}	2.00	1.8×10^{-4}
4.0×10^{-3}	3.0×10^{-3}	1.00	7.2×10^{-4}

Is the predicted rate law expression derived from the mechanism consistent with these data?

We are looking for the rate expression for:

$$I^-_{aq} + OCl^-_{aq} \longrightarrow OI^-_{aq} + Cl^-_{aq}$$

$$R = \frac{d[OI^-]}{dt} = k_3[OI^-][HOI]$$

Using the steady-state approximation for the intermediates $[HOI]$ and $[HOCl]$ yields:

$$\frac{d[HOI]}{dt} = 0 = -k_3[HOI][OH^-] + k_2[I^-][HOCl]$$

$$[HOI] = \frac{k_2}{k_3}\frac{[I^-][HOCl]}{[OH^-]}$$

$$\frac{d[HOCl]}{dt} = 0 = -k_2[I^-][HOCl] + k_{-1}[HOCl][OH^-] + k_1[OCl^-][H_2O]$$

$$k_2[I^-][HOCl] - k_{-1}[HOCl][OH^-] = k_1[OCl^-][H_2O]$$

$$[HOCl](k_2[I^-] - k_{-1}[OH^-]) = k_1[OCl^-][H_2O]$$

$$[HOCl] = \frac{k_1[OCl^-][H_2O]}{(k_2[I^-] - k_{-1}[OH^-])}$$

Combining yields:

$$R = \frac{d[OI^-]}{dt} = k_3[OI^-][HOI] = k_3[OI^-]\frac{k_2}{k_3}\frac{[I^-][HOCl]}{[OH^-]}$$

$$= k_2[OI^-]\frac{[I^-][HOCl]}{[OH^-]} = k_2[OI^-]\frac{[I^-]}{[OH^-]}\frac{k_1[OCl^-][H_2O]}{(k_2[I^-] - k_{-1}[OH^-])}$$

$$= k_2 k_1 \frac{[I^-][OI^-]}{[OH^-]}\frac{[OCl^-][H_2O]}{(k_2[I^-] - k_{-1}[OH^-])}$$

P26.8) Consider the following mechanism that describes the formation of product *P*:

$$A \underset{k_{-1}}{\overset{k_1}{\rightleftarrows}} B \underset{k_{-2}}{\overset{k_2}{\rightleftarrows}} C$$

$$B \xrightarrow{k_3} P$$

If only the species A is present at $t = 0$, what is the expression for the concentration of P as a function of time? You can apply the preequilibrium approximation in deriving your answer.

Using the preequilibrium approximation we can write:

$$\frac{d[P]}{dt} = k_3[B]$$

$$[B] = \frac{k_1}{k_{-1}}[A] + \frac{k_{-1}}{k_1}[C]$$

$$\frac{d[P]}{dt} = k_3[B] = k_3\left(\frac{k_1}{k_{-1}}[A] + \frac{k_{-1}}{k_1}[C]\right) = k_3\left(K_{C1}[A] + K_{C2}[C]\right), \text{ with } K_{C1} = \frac{k_1}{k_{-1}}, \text{ and } K_{C2}\frac{k_{-1}}{k_1}$$

P26.9) The enzyme fumarase catalyzes the hydrolysis of fumarate: Fumarate + $H_2O \rightarrow$ L-malate. The turnover number for this enzyme is $2.5 \times 10^3\ s^{-1}$, and the Michaelis constant is 4.2×10^{-6} M. What is the rate of fumarate conversion if the initial enzyme concentration is 1×10^{-6} M and the fumarate concentration is 2×10^{-4} M?

Using the Michaelis-Menten equation gives for the rate of conversion:

$$R_0 = \frac{k_2[S_0][E_0]}{([S_0]+K_m)} = \frac{(2.5\times10^3\ s^{-1})\times(2\times10^{-4}\ M)\times(1\times10^{-6}\ M)}{((2\times10^{-4}\ M)+(4.2\times10^{-6}\ M))} = \underline{2.45\times10^{-3}\ s^{-1}}$$

P26.14) The enzyme glycogen synthase kinase 3β (GSK–3β) plays a central role in Alzheimer's disease. The onset of Alzheimer's disease is accompanied by the production of highly phosphorylated forms of a protein referred to as τ. GSK–3β contributes to the hyperphosphorylation of τ such that inhibiting the activity of this enzyme represents a pathway for the development of an Alzheimer's drug. A compound known as Ro 31-8220 is a competitive inhibitor of GSK-3β. The following data were obtained for the rate of GSK-3β activity in the presence and absence of Ro 31-8220 [A. Martinez *et al., J. Medicinal Chemistry* 45 (2002), 1292]:

	***Rate$_0$* (μM s^{-1}),**	***Rate$_0$* (μM s^{-1})**
[S] (μM)	**[I] = 0**	**[I] = 200 μM**
66.7	4.17×10^{-8}	3.33×10^{-8}
40.0	3.97×10^{-8}	2.98×10^{-8}
20.0	3.62×10^{-8}	2.38×10^{-8}
13.3	3.27×10^{-8}	1.81×10^{-8}
10.0	2.98×10^{-8}	1.39×10^{-8}
6.67	2.31×10^{-8}	1.04×10^{-8}

Determine K_m and R_{max} for GSK–3β and, using the data with the inhibitor, determine K_m^* and K_i.

The Lineweaver-Burk plot for the data with (red) and without (black) an inhibitor:

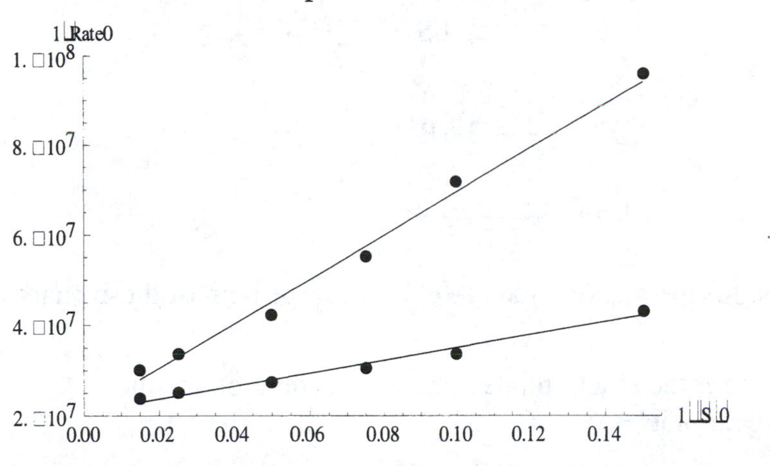

From the plot without an inhibitor we obtain R_{max} and K_m:

$$\text{slope} = \frac{\Delta(1/\text{Rate}_0)}{\Delta(1/[\text{S}]_0)} = \frac{\left(\left(4.329\times10^7\ \text{s}\,\mu\text{M}^{-1}\right)-\left(2.39808\times10^7\ \text{s}\,\mu\text{M}^{-1}\right)\right)}{\left(\left(0.149925\,\mu\text{M}^{-1}\right)-\left(0.0149925\,\mu\text{M}^{-1}\right)\right)} = 1.43103\times10^8\ \text{s}$$

$$\text{inter} = \frac{1}{\text{R}_0} - \frac{\text{slope}}{[\text{S}]_0} = \frac{1}{\left(4.17\times10^{-8}\ \mu\text{M s}^{-1}\right)} - \frac{\left(1.43103\times10^8\ \text{s}\right)}{(66.7\ \mu\text{M})} = 2.18353\times10^7\ \text{s}\,\mu\text{M}^{-1}$$

$$\text{R}_{\text{max}} = \frac{1}{\text{inter}} = \frac{1}{\left(2.18353\times10^7\ \text{s}\,\mu\text{M}^{-1}\right)} = \underline{4.58\times10^{-8}\ \mu\text{M s}^{-1}}$$

$$\text{K}_\text{m} = \text{slope}\,\text{R}_{\text{max}} = \left(1.43103\times10^8\ \text{s}\right)\times\left(4.58\times10^{-8}\ \mu\text{M s}^{-1}\right) = \underline{6.55\ \mu\text{M}}$$

From the plot with an inhibitor we obtain and K_m*:

$$\text{slope} = \frac{\Delta(1/\text{Rate}_0)}{\Delta(1/[\text{S}]_0)} = \frac{\left(\left(9.61538\times10^7\ \text{s}\,\mu\text{M}^{-1}\right)-\left(3.003\times10^7\ \text{s}\,\mu\text{M}^{-1}\right)\right)}{\left(\left(0.149925\,\mu\text{M}^{-1}\right)-\left(0.0149925\,\mu\text{M}^{-1}\right)\right)} = 4.90051\times10^8\ \text{s}$$

$$\text{K}_\text{m}^* = \text{slope}\,\text{R}_{\text{max}} = \left(4.90051\times10^8\ \text{s}\right)\times\left(4.58\times10^{-8}\ \mu\text{M s}^{-1}\right) = \underline{22.44\ \mu\text{M}}$$

And finally K_i:

$$\text{K}_\text{m}^* = \text{K}_\text{m}\left(1+\frac{[\text{I}]}{\text{K}_\text{i}}\right)$$

$$\text{K}_\text{i} = \frac{[\text{I}]}{\left(\dfrac{\text{K}_\text{m}^*}{\text{K}_\text{m}}-1\right)} = \frac{\left(200\ \mu\text{M}^{-1}\right)}{\left(\dfrac{(22.44\ \mu\text{M})}{(6.55\ \mu\text{M})}-1\right)} = \underline{82.4\ \mu\text{M}^{-1}}$$

P26.18) In addition to competitive and noncompetitive enzyme inhibition, inhibition can also be *un*competitive. This occurs when binding of the inhibitor can occur only after the substrate is bound. The mechanism for this type of inhibition is:

$$E + S \underset{k_{-1}}{\overset{k_1}{\rightleftarrows}} ES$$

$$ES \xrightarrow{k_2} E + P$$

$$ES + I \underset{k_{-4}}{\overset{k_4}{\rightleftarrows}} EIS$$

a. Derive the expression for the reaction rate. Are K_m, R_{max}, or both of these quantities affected?

b. Using your expression for the reaction rate, what is the equation for the corresponding reciprocal plot?

c. How does the reciprocal plot for uncompetitive inhibition deviate from those of competitive and noncompetitive inhibition? That is, could you still use reciprocal plots at varying inhibitor concentrations to differentiate between these inhibition mechanisms?

a) For the noncompetitive enzyme inhibition we obtain:

$$[E]_0 = [E] + [ES] + [EIS]$$

$$[E]_0 = \frac{K_m [ES]}{[S]} + [ES] + \frac{[I][ES]}{K_i}$$

$$[E]_0 = [ES]\left(\frac{K_m}{[S]} + 1 + \frac{[I]}{K_i}\right)$$

$$[ES] = \frac{[E]_0}{\left(\frac{K_m}{[S]} + \frac{[I]}{K_i} + 1\right)}$$

$$R_0 = \frac{d[P]}{dt} = k_2 [ES] = \frac{k_2 [E]_0}{\left(\frac{K_m}{[S]} + \frac{[I]}{K_i} + 1\right)} = \frac{R_{max}}{\left(\frac{K_m}{[S]} + \frac{[I]}{K_i} + 1\right)}$$

With the assumption $[S] \cong [S]_0$:

$$R_0 = \frac{[S]_0 R_{max}}{\left(K_m + [S]_0\left(\frac{[I]}{K_i} + 1\right)\right)}$$

b) The reciprocal plot is then:

$$\frac{1}{R_0} = \frac{\left(K_m + [S]_0\left(\frac{[I]}{K_i}+1\right)\right)}{[S]_0 R_{max}} = \frac{K_m}{[S]_0 R_{max}} + \frac{[S]_0\left(\frac{[I]}{K_i}+1\right)}{[S]_0 R_{max}} = \frac{K_m}{[S]_0 R_{max}} + \frac{\left(\frac{[I]}{K_i}+1\right)}{R_{max}}$$

$$\frac{1}{R_0} = \frac{1}{[S]_0} + \frac{1}{R_{max}}\left(\frac{[I]}{K_i}+1\right)$$

c) A plot of $\frac{1}{R_0}$ versus $\frac{1}{[S]_0}$ yields:

$$\text{Slope} = \frac{K_m}{R_{max}}$$

$$\text{Intercept} = \frac{1}{R_{max}}\left(\frac{[I]}{K_i}+1\right)$$

P26.22) If $\tau_f = 1 \times 10^{-10}$ s and $k_{ic} = 5 \times 10^8\ \text{s}^{-1}$, what is ϕ_f? Assume that the rate constants for intersystem crossing and quenching are sufficiently small that these processes can be neglected.

Φ_f of the process is given by:

$$\Phi_f = 1-(\tau_f\ k_{ic}) = 1-(1\text{x}10^{-10}\text{s})\times(5\times10^8\text{s}^{-1})\ \underline{=0.95}$$

P26.24) If 10% of the energy of a 100-W incandescent bulb is in the form of visible light having an average wavelength of 600 nm, how many quanta of light are emitted per second from the lightbulb?

We first calculate the energy of one photon with a wave length of 600 nm:

$$E_{photon} = \frac{h\,c}{\lambda} = \frac{(6.626\times10^{-34}\ \text{J s})\times(2.998\times10^8\ \text{m s}^{-1})}{(600\times10^{-9}\ \text{m})} = 3.31079\times10^{-19}\ \text{J}$$

The number of quanta is then:

$$n_{quanta} = \frac{(0.1)\times(100\ \text{J s}^{-1})}{(3.31079\times10^{-19}\ \text{J})}\ \underline{= 3.02\times10^{19}}$$

P26.27) A central issue in the design of aircraft is improving the lift of aircraft wings. To assist in the design of more efficient wings, wind-tunnel tests are performed in which the pressures at various parts of the wing are measured generally using only a few localized pressure sensors. Recently, pressure-sensitive paints have been developed to provide a more detailed view of wing pressure. In these paints, a luminescent molecule is dispersed into an oxygen-permeable paint and the aircraft wing is painted. The wing is placed into an airfoil, and luminescence from the paint is measured. The variation in O_2 pressure is measured by monitoring the luminescence intensity, with lower intensity demonstrating areas of higher O_2 pressure due to quenching.

a. The use of platinum octaethylprophyrin (PtOEP) as an oxygen sensor in pressure-sensitive paints was described by Gouterman and coworkers [*Review of Scientific Instruments* 61 (1990), 3340]. In this work, the following relationship between luminescence intensity and pressure was derived: $I_0/I = A + B(P/P_0)$ where I_0 is the fluorescence intensity at ambient pressure P_0, and I is the fluorescence intensity at an arbitrary pressure P. Determine coefficients A and B in the preceding expression using the Stern–Volmer equation: $k_{total} = 1/\tau_l = k_r + k_q[Q]$. In this equation τ_l is the luminescence lifetime, k_r is the luminescent rate constant, and k_q is the quenching rate constant. In addition, the luminescent intensity ratio is equal to the ratio of luminescence quantum yields at ambient pressure, Φ_0, and an arbitrary pressure, Φ:

$$\Phi_0/\Phi = I_0/I.$$

b. Using the following calibration data of the intensity ratio versus pressure observed for PtOEP, determine A and B:

I_0/I	P/P_0	I_0/I	P/P_0
1.0	1.0	0.65	0.46
0.9	0.86	0.61	0.40
0.87	0.80	0.55	0.34
0.83	0.75	0.50	0.28
0.77	0.65	0.46	0.20
0.70	0.53	0.35	0.10

c. At an ambient pressure of 1 atm, $I_0 = 50{,}000$ (arbitrary units) and 40,000 at the front and back of the wing. The wind tunnel is turned on to a speed of Mach 0.36 and the measured luminescence intensity is 65,000 and 45,000 at the respective locations. What is the pressure differential between the front and back of the wing?

a) Based on:

$$\frac{I_0}{I} = A + B\frac{p}{p_0}$$

we need to plot $\frac{I_0}{I}$ versus $\frac{p}{p_0}$ to obtain the constants A and B:

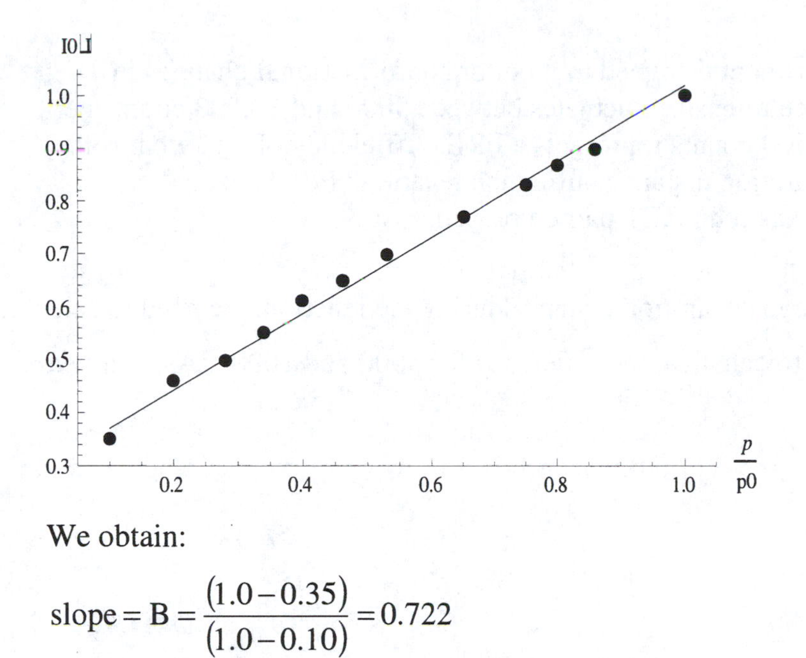

We obtain:

$$\text{slope} = B = \frac{(1.0-0.35)}{(1.0-0.10)} = 0.722$$

$$A = \frac{I_0}{I} - B\frac{p}{p_0} = (0.1)-(0.722)\times(1.0) \quad \underline{= 0.278}$$

(Also: A + B =1)

b) We can use the results from a) to calculate the pressures at the front and back at the wing:

$$\frac{I_0}{I} = A + B\frac{p}{p_0}$$

$$p_{front} = \frac{\left(\frac{I_{0,front}}{I_{front}} - A\right)}{B} p_0 = \frac{\left(\frac{(50000)}{(65000)} - (0.278)\right)}{(0.722)} \times (1\ \text{atm}) = 0.680\ \text{atm}$$

$$p_{back} = \frac{\left(\frac{I_{0,back}}{I_{bak}} - A\right)}{B} p_0 = \frac{\left(\frac{(40000)}{(45000)} - (0.278)\right)}{(0.722)} \times (1\ \text{atm}) = 0.846\ \text{atm}$$

And the difference:

$$\Delta p = p_{front} - p_{front} = (0.680\ \text{atm}) - (0.846\ \text{atm}) \quad \underline{= -0.166\ \text{atm}}$$

P26.30) In a FRET experiment designed to monitor conformational changes in T4 lysozyme, the fluorescence intensity fluctuates between 5000 and 10,000 counts per section. Assuming that 7500 counts represents a FRET efficiency of 0.5, what is the change in FRET pair separation distance during the reaction? For the tetramethylrhodamine/Texas red FRET pair employed, $r_0 = 50$ Å.

To determine the increase in separation distance during the reaction, we need to calculate r_{5000} and r_{10000}. We need to calculate the efficiency for 5000 and 10000. Assuming a linear relationship

$$Eff(7500) = 0.5$$

$$\therefore \frac{Eff(5000)}{Eff(7500)} = \frac{X}{0.5}$$

$$Eff(5000) = 0.333$$

$$Eff(1000) = 0.667$$

$$\Delta r = r_{5000} - r_{10000}$$

$$r^6{}_{5000} = \frac{r_0^{\,6}}{Eff(5000)} - r_0^{\,6}$$

$$r^6 = \frac{(50\,\overset{\circ}{A})^6}{0.333} + (50\,\overset{\circ}{A})^6$$

$$r = 56.05\,\overset{\circ}{A}$$

$$r^6{}_{10000} = \frac{r_0^{\,6}}{Eff(10000)} - r_0^{\,6}$$

$$r^6 = \frac{(50\,\overset{\circ}{A})^6}{0.667} + (50\,\overset{\circ}{A})^6$$

$$r = 44.5\overset{\circ}{A}$$

$$\Delta r = r(5000) - r(10000)$$

$$= 56.05 - 44.54\ \overset{\circ}{A} = \underline{11.5\ \overset{\circ}{A}}$$

P26.32) In Marcus theory for electron transfer, the reorganization energy is partitioned into solvent and solute contributions. Modeling the solvent as a dielectric continuum, the solvent reorganization energy is given by

$$\lambda_{sol} = \frac{(\Delta e)^2}{4\pi\varepsilon_0}\left(\frac{1}{d_1}+\frac{1}{d_2}-\frac{1}{r}\right)\left(\frac{1}{n^2}-\frac{1}{\varepsilon}\right)$$

Where Δe is the amount of charge transferred, d_1 and d_2 are the ionic diameters of ionic products, r is the separation distance of the reactants, n^2 is the index of refraction of the surrounding medium, and ε is the dielectric constant of the medium. In addition, $(4\pi\varepsilon_0)^{-1} = 8.99 \times 10^9$ J m C^{-2}.

a. For an electron transfer in water (n = 1.33 and ε = 80), for which the ionic diameters of both species are 6 Å and the separation distance is 15 Å, what is the expected solvent reorganization energy?

b. Redo the above calculation for the the same reaction occurring in a protein. The dielectric constant of a protein is dependent on sequence, structure, and the amount on included water; however, a dielectric constant of 4 is generally assumed consistent with a hydrophobic environment.

We use the equation for the solvent reorganization energy from the Marcus theory:

$$\lambda_{sol} = \frac{\Delta e^2}{4\pi\,\varepsilon_0}\left(\frac{1}{d_1}+\frac{1}{d_2}-\frac{1}{r}\right)\left(\frac{1}{n^2}-\frac{1}{\varepsilon}\right)$$

a) With the transfer of one electron $(\Delta e = 1)$, and $\varepsilon = 80$ we obtain:

$$\lambda_{sol} = \left(1.602\times10^{-19}\ \mathrm{C}\right)^2 \times \left(8.99\times10^{9}\ \mathrm{J\ m\ C^{-2}}\right)$$

$$\times\left(\frac{1}{(6\times10^{-10}\ \mathrm{m})}+\frac{1}{(6\times10^{-10}\ \mathrm{m})}-\frac{1}{(15\times10^{-10}\ \mathrm{m})}\right)\times\left(\frac{1}{(1.33)^2}-\frac{1}{(80)}\right) = \underline{3.40\times10^{-19}\ \mathrm{J}}$$

b) With the transfer of one electron $(\Delta e = 1)$, and $\varepsilon = 4$ we obtain:

$$\lambda_{sol} = \left(1.602\times10^{-19}\ \mathrm{C}\right)^2 \times \left(8.99\times10^{9}\ \mathrm{J\ m\ C^{-2}}\right)$$

$$\times\left(\frac{1}{(6\times10^{-10}\ \mathrm{m})}+\frac{1}{(6\times10^{-10}\ \mathrm{m})}-\frac{1}{(15\times10^{-10}\ \mathrm{m})}\right)\times\left(\frac{1}{(1.33)^2}-\frac{1}{(4)}\right) = \underline{1.94\times10^{-19}\ \mathrm{J}}$$